INSANE MODE

HOW ELON MUSK'S TESLA
SPARKED AN ELECTRIC REVOLUTION
TO END THE AGE OF OIL

HAMISH McKENZIE

DUTTON

DUTTON

An imprint of Penguin Random House LLC
penguinrandomhouse.com

Previously published as a Dutton hardcover edition in 2018

First trade paperback printing: October 2019

THE LIBRARY OF CONGRESS HAS CATALOGUED THE HARDCOVER EDITION AS FOLLOWS:
Names: McKenzie, Hamish, author.
Title: Insane mode : how Elon Musk's Tesla sparked an electric revolution
to end the age of oil / Hamish McKenzie.
Description: New York : Dutton, [2017] | Includes index.
Identifiers: LCCN 2016057562 (print) | LCCN 2017031039 (ebook) |
ISBN 9781101985977 (ebook) | ISBN 9781101985953 (hardcover)
Subjects: LCSH: Musk, Elon. | Tesla Motors. | Electric vehicle industry—United States. |
Alternative fuel vehicle industry—United States. | Electric power. | Renewable energy sources.
Classification: LCC HD9710.U54 (ebook) | LCC HD9710.U54 T4763 2017 (print) |
DDC 338.7/6292293092—dc23
LC record available at https://lccn.loc.gov/2016057562

Dutton trade paperback ISBN: 9781101985960

Printed in the United States of America
1 3 5 7 9 10 8 6 4 2

Set in Adobe Garamond

For Steph, for being there

CONTENTS

PART THREE: THE OPEN ROAD

PART ONE

INDUCTION

1

GET YOUR MOTOR RUNNIN'

"In certain sectors like automotive and solar and space,
you don't see new entrants."

The first car I drove for any reasonable period of time was a 1983 Ford Laser with a manual choke. As a sixteen-year-old who needed to get places, I learned the delicate art of gradually modulating the choke to achieve the perfect mixture of air and gasoline so that the little Laser would purr like a panther in a piano box. The car's paint was gold, but the years had faded its luster so that it settled into more of a dusky brown. I called it Brown-Brown and drove it all around Alexandra, New Zealand—population 5,000—and to nearby swimming holes, sports grounds, and make-out spots in the scrubby hills that surrounded my hometown.

Other than mastering the choke, I didn't know much about the car and didn't really care to find out. My dad, a physicist who knew how to choreograph Brown-Brown's bits and bobs so that it performed the miracle of propulsion, took care of all the maintenance. All I had to do was fill it up with gas and stop it from stalling on a black-iced back road in the middle of nowhere. And that was fine with me.

Later, while I was picking fruit at a local orchard during university holidays to earn rent money, I did make an attempt to learn how cars worked. By that point, I had upgraded to a 1991 Toyota Corona, which by my standards was a luxury vehicle. It was not only chokeless but it also had an automatic transmission. One hot day, I was on the top step of my ladder among cherry trees while my car-literate friend in the neighboring tree explained to me how an internal combustion engine works. Despite my father's influence—and much to his disappointment—I was an arts student and did not have a mind for mechanics. While I committed terms like *carburetor, piston*, and *camshaft* to memory between mouthfuls of cherries, I struggled to recall in which order they interacted, or if they interacted at all. My friend soon grew frustrated with my ineptitude, and I resigned myself to the notion that this fiendishly complicated wizardry would remain forever out of my reach. And that was fine with me.

My ambivalent relationship with motor vehicles continued even after, at twenty-nine years old, I moved to the United States of America, the spiritual home of the automobile. At the wheel of my wife's 2001 Honda Civic, I learned how to drive on the wrong side of the road and fine-tuned my aggression on the gas pedal so that I could stave off death on the highways, but I remained ignorant of how spark plugs sparked and timing belts belted. Indeed, I avoided driving whenever I could and came to believe that the world would be better off without cars. In one of the first pieces I wrote since joining the tech news site *PandoDaily*, I implored Silicon Valley to rid us of them. I felt that the environmental costs of cars and roads were unacceptable when the climate was warming at such a rate that there'd soon be more deaths from heatstroke than from motor accidents. Cars were death traps, health hazards, planet killers, and insidious isolation engines, I reasoned. Who'd want them?

Of course, lots of people wanted them, and path dependency is real. We'd already carved up mountains, paved over swamplands, and invented garages to cater to our four-wheeled wonder wagons, so giving up on them now hardly seemed realistic. After a multitude of commenters disabused me of my car-free fantasy, I breathed a sigh of concession and moved on.

It was about then that I discovered Tesla.

I had joined *Pando* in April 2012, a few months after Steve Jobs, the cofounder and CEO of Apple, died, and I found a tech world still grieving the loss of its superstar. The industry was bereft of a figure who could command the world's attention with the twitch of a stage-managed eyebrow, a man who could send the media into conniptions with an addendum to a slide show. Silicon Valley was frantically searching for one more thing, but results had been mixed. The iPhone was by then status quo and the Great Innovators of the Valley had turned their attention to photo-sharing apps and ad optimization. Software engineers were earning millions to digitize aggregated attention and make it amenable to the distribution of newsfeed flyers. Other ideas failed to inspire. Facebook, but for small groups of people? Limos on demand, but for middle-class San Franciscans? Marissa Mayer, but for Yahoo!?

Then, in June 2012, the Tesla Model S came along. While it enjoyed a splendid launch party, the public didn't know much about it at first. The luxury electric sedan came with a $70,000 price tag, and that was just for the cheapest version. At the launch event, Tesla handed over the keys to only ten cars, with plans to scale up production later. Reviewers got ten-minute test drives. Still, it was enough to capture the imaginations of the auto and tech media. *The Wall Street Journal*'s Dan Neil compared the Model S to a Lamborghini and praised the marvel of its silent ride. *Wired* said it was "a complete hoot to drive." The performance version of the car accelerated from zero to sixty miles per hour in 4.2 seconds. That was supercar territory—in a sedan.

The next month, Tesla's CEO, Elon Musk, appeared at the Pando-Monthly speaker series in San Francisco. I was in China at the time, but I watched a video of the event online. I knew little of Musk but was instantly struck by his plainspoken audacity. He already had a rocket company, SpaceX, that sent payloads to the International Space Station, and he had conceived the solar power start-up SolarCity, which he also funded. With Tesla, he was intent on weaning the world off fossil fuels. "I'm trying to allocate my efforts to that which I think would most affect the future of humanity in a positive way," he told my then boss, Sarah Lacy, at the event. "There's lots of entrepreneurial energy and financing heading towards the Internet, whereas in certain sectors like automotive and solar and space, you don't see new entrants."

If we were going to be stuck with cars, I figured, we might as well let this guy make them electric so we can at least stop pumping so much carbon dioxide into the atmosphere.

In reading more about Tesla, I found that it had launched an electric sports car, the Roadster, in 2008. It was the first cool electric car, the first demonstration that a vehicle powered by an electric motor was more interesting than a golf cart. With a price tag in the $100,000 range, it was sold largely to rich people and celebrities, which was a not bad way to win attention but also, because of the expense of the battery, an economic necessity. Musk, however, had started talking about a fully electric family car in 2008, and it had taken a while to eventuate. I wondered why. Then I watched *Revenge of the Electric Car*, a 2011 documentary that showed Tesla struggling to survive the financial crisis. I read news stories and magazine profiles that told of how Musk paid Tesla employees out of his own pocket to keep the company alive. Tesla was on bankruptcy's doorstep at the end of 2008, before it was saved at the last minute by a $40 million investment and then, the next year, a helping hand from Daimler. Over the following years, it bought a factory, went public, and then created the Model S, which went on to win *Motor Trend*'s Car of the Year award—the first unanimous winner in the magazine's history. Maybe this Musk guy was onto something.

By the middle of 2013, Tesla's stock price had shot above $160 and its market valuation approached $20 billion. Mom-and-pop investors who had bought the stock for around $20 a share in 2010 became millionaires. Musk started to get famous—not just in the tech world but in the real world, too. In August 2013, his notoriety reached a new level when he announced plans for a "fifth mode of transport" that he said could take passengers from Los Angeles to San Francisco in half an hour. He wrote the blueprint for his so-called Hyperloop in an all-nighter and then published it on the Tesla and SpaceX corporate blogs. He didn't plan to build the Hyperloop himself, but he hoped someone else would make it a reality. The ensuing news coverage bestowed on Musk the kind of attention usually reserved for Steve Jobs.

Given the task of coming up with an article about the Hyperloop announcement for *Pando*, I wrote that Musk was more important to society than Jobs ever was. While Jobs did the world a great service by putting powerful Internet-connected computers in our pockets, Musk was operating on a different plane of purpose. In attempting to transform transportation and radically improve space travel instead of developing another photo-sharing app or the next *Flappy Bird*, Musk set an example for a new generation of entrepreneurs.

After that piece ran, a nonfiction editor e-mailed to ask if I would be interested in writing a book about Musk. Reading the e-mail while dressed in boxers and a T-shirt in the spare bedroom that doubled as an office in my Baltimore apartment, I mulled the suggestion and concluded that, actually, yes, it was a good idea. I took the proposal to Musk but was surprised when he instead offered me a job at Tesla. After some hesitation—I was not eager to leave journalism—I ultimately accepted. After all, I thought, I could always come back to the book.

I spent just over a year at Tesla but discovered that journalism was an itch I hadn't finished scratching. I left in March 2015 and, indeed, came back to the book. Read this book with these caveats, then: Yes, I'm a former Tesla employee. I believe in the company's mission. I even hold Tesla stock. But I am also committed to serving the reader. In these pages, I strive to present a fair and clear-eyed view of what's great about Tesla, and of the very real challenges it faces.

This book, however, isn't an insider's tale—I will leave that work to the gossip blogs—and it's not just about Tesla. It's about something much bigger. It's a story about how one determined Silicon Valley start-up changed the entire auto industry, along the way inspiring a slew of well-funded imitators from California to China. It's a system-level view of a technological and economic transformation that will affect the lives of everyone on the planet. It is the story of a revolution that Tesla started.

When I first drove the Tesla Model S, I thought of it as a computer on wheels. Its digital controls, Internet connection, software updates, and iPad-like touch screen do tend to create that impression. But that description undersells its promise. The Model S—like all of Tesla's cars—can be better thought of as a battery on wheels. Just look at it. Stripped of its shell and seats, the machine is essentially a set of four wheels bracing a low-slung metallic mattress that contains several thousand cylindrical lithium-ion batteries like those used in old laptops. Peel off the lid and you'll see the batteries standing on end and packed rump to rump in eight modules, arrayed in tidy rows like obedient schoolchildren. It is this modest configuration of cells that is finally bringing an end to the oil industry's dominance of global energy supply.

Tesla is a vehicle for an idea: that we humans have better ways to power our lives than to burn a dinosaur-era compaction that dirties the air and skanks up the chemistry of the atmosphere. That notion applies to more than just cars. Tesla also sells its batteries as energy storage units. Since it acquired SolarCity in 2016 and added solar panels to its offerings, Musk has made his intentions clear: Tesla is an energy company.

This is the story of how the electric car became a Trojan horse for a new energy economy. I believe it is the most important technology story of the twenty-first century. And it finally inspired me to figure out, once and for all, how an internal combustion engine works—just in time for it to disappear.

2

A RUSH OF ELECTRONS
TO THE HEAD

"Your own personal roller coaster."

In the summer of 2014, my father came to visit me in San Francisco from New Zealand. To treat him, I borrowed a Model S for the weekend. I didn't tell him that I had it, but soon after he arrived, I suggested that we take a walk to a nearby park, where I had parked the car. As we approached, I feigned surprise, pointed across the street, and said, "Oh, look, there's a Model S!" Dad, a sixty-four-year-old Elon Musk fanboy who had never seen a Tesla in person, immediately walked over. As he cupped his hands against the front window and peered inside, I walked up behind him and surreptitiously clicked the key fob I had secreted in my pocket. The chrome door handles responded by automatically extending. Dad stepped back in surprise. "Let's get in," I said. He laughed with the delight of a child.

The next day, we took the Model S to Napa Valley, where we visited vineyards with friends, who gushed over the slick red sedan. "You know you've made it when you're driving around Napa in a Tesla!"

one enthused. By mid-2014, two years since it first hit the road, Tesla's Model S had attained popular status as a kind of fetish item for people impressed by the latest gadgets or material indicators of wealth. The car's auto-retracting door handles gave it a signature feature and provided an immediate conversation point. It looked good enough to blend in at even the most upscale Napa resorts. And people familiar with Tesla instantly recognized the car as a symbol of Silicon Valley innovation, of forward thinking, as a step out of the fossil fuels era.

On Napa's back roads, I gave Dad a turn in the driver's seat. I had been driving semi-cautiously most of the day to preserve the car's range. It's about sixty miles to Napa from San Francisco, and I wanted to be sure that we had enough juice in the battery to get us there and back comfortably, while taking into account the extra miles required to tour the wineries. At that time, the nearest charging station was forty miles away in the wrong direction. But I couldn't deny Dad the pure pleasure of burning rubber in the most consequential automobile since the Model T.

The Model S was the first car Tesla had produced entirely on its own and it was the first car to provide any hint that the era of the internal combustion engine's dominance could be coming to an end. A single charge of its eighty-five kilowatt-hour battery gave the car the ability to drive 265 miles. For the first time, an electric-car owner could drive far out of town and be confident of returning home without running out of juice. It boasted impressive high-tech credentials, including a seventeen-inch touch screen that served as central command, allowing occupants to access maps, control the sound system, and retract the sunroof. Improvements, such as automated ride heights and creep control, could be delivered via over-the-air software updates, as if it were a laptop computer. And drivers could refuel the car for free at high-speed charging stations—Superchargers—that Tesla was distributing around the world.

Unlike its electric predecessors, such as the Nissan Leaf and the

Mitsubishi i-MiEV, the Model S was also exceedingly practical, able to accommodate seven passengers—counting the two optional rear-facing seats—and offering more than sixty-three cubic feet of storage, including a front trunk that took advantage of space freed up from the absence of an engine block. While it had an aluminum shell and sat atop a lithium-ion battery that could, without thermal protection, combust in spectacular fashion, the Model S also rated well on safety. Its thousand-pound battery pack was laid flat and integrated into the chassis beneath the passenger compartment, so the car had a low center of gravity that made it difficult to roll. Without an engine block, the front of the vehicle had more crumple room to absorb the energy of a collision, and the roof, reinforced with extruded aluminum and boron steel, broke the machine that was testing its strength.

With a top-line price of about $100,000, the car was far from cheap, but it quickly attained a cult status, particularly among the wealthy tech class in California, where Tesla found its early adopters. Like Apple's iPod, it was a beautiful and useful consumer item that, while expensive, made its competitors look quaint. At the end of 2012, it was given almost all the awards the auto industry had to offer, the most prominent of which was *Motor Trend*'s Car of the Year. But, most importantly, the Model S was a blast to drive. The car's electric motors produced torque instantly, allowing it to reach highway speed in about four seconds. A stomp on the accelerator delivered a roller-coaster rush.

As Dad guided our 4,647-pound aluminum steed around a bend and onto a strip of open road, I urged him to let rip. You can imagine what the next moment would look like in slow motion, as if it were a scene from *Fast & Furious: Grandpa's Revenge*. The camera would zoom in close on his tattered right sneaker as it lifted off the accelerator to maximize leverage for the impending pedal-punch. The background music would warp and blur into a chewed-tape version of a power-rock song while the universe sucked in a deep breath. Then, on its downswing, the seven-year-old sneaker would move with achingly slow surety to its rubber-clad destiny before finally unleashing the fury of the leg's pent-up force on the unsuspecting pedal. At this point, the tape would return to normal speed as the soundtrack's power chords burst into adrenaline-spilling comprehensibility, and the pedal would be slammed without apology to the carpet. The cameras would promptly cut to our upper extremities to show our heads being whipped back against the headrests, our stomachs getting sucked into a flatness they hadn't known since we were teenagers, and the stunned, stupid grins that assaulted our faces. Such is the result of a Model S suddenly summoning a torrent of electrons from its battery pack. It's what zero to sixty miles per hour in 4.2 seconds feels like.

"Not bad," Dad said.

The Model S is so quick because its electric induction motor can deliver maximum torque from standstill. That same motor is also able to draw on power faster than a conventional car, simply because electrons travel faster from battery to motor than gasoline does from gas tank to piston. The car also has instant access to a tremendous amount of horsepower—the Model S we were driving had 416 horsepower, which is comparable to a Ford Mustang—and it doesn't have to deal with the acceleration lags that come with shifting from first gear into second, second into third, and so on. The car can just keep accelerating smoothly until it hits top speed. In fact, perhaps the main impediment to even quicker acceleration is its tires, which would start slipping and billowing smoke if they were forced to spin any faster. Finally, the low, heavy battery pack helps keep the car balanced, so there's even pressure on all the contact points on the road. That helps the car stick to the road like burnt hash browns to a pan.

By contrast, a gasoline car requires a large number of steps to convert potential energy from the fuel into motion. The vehicle can't get started without fuel injectors delivering bursts of fuel into the engine (or, in older cars, pumping fuel and air into the carburetor), where the fuel and air are mixed in the necessary proportions for combustion. A spark plug ignites the mixture to cause an explosion that drives the piston down, creating torque that eventually turns the wheels. For all those things to happen, the engine needs to already be turning, which requires an electric starter motor powered by a twelve-volt battery. Some of the mechanical energy from the motor is diverted to an alternator, which keeps the battery fully charged. Meanwhile, as it accelerates, the car has to keep shifting gears upward to reach cruising speed. The gears are necessary because the motor's torque output can be maintained over only a small range of engine rotation speeds. Complicating matters further, the shape of the car can act like a plane's wing—air passes over the car, but it takes a longer path than air that travels under the car. The resulting lower pressure above the car puts it in a constant fight with gravity, so its natural inclination at speed is to lift itself *off* the road. This is a high-speed issue for both conventional and electric cars, but the extra-low weight in the Tesla mitigates the problem. For gasoline cars, the lumpy weight distribution that comes from having a heavy engine block sitting high in the front or back of the vehicle adds an extra challenge in getting the car to stick to the road, especially on corners.

By this point, you might be thinking that I'm some kind of propagandist for electric cars. And yeah, maybe there is bias at work here. But it's high time that some bias cut in favor of electric cars. It has, after all, been about twelve decades. Consider the reasons we've been given over the last 120 years for why electric cars just aren't right for this world:

> **Cost:** The high cost of the batteries needed to power electric cars makes them uneconomical. The Nissan Leaf, for instance, was more expensive than a Nissan Versa but could drive only a quarter of the distance between refueling stops and had comparable performance.

Range: Before the Tesla Roadster, the range of commercially available electric cars limited them to short trips.

Refueling time: It takes only a few minutes to stick a hose into your gas tank and fill it to the brim. Most electric cars, however, need hours to fully charge.

Infrastructure: Gas stations are everywhere, so you seldom have to worry about running out of fuel while on a long-distance trip. Electric cars, on the other hand, need charging stations for long trips, and they are still in relatively short supply.

Cold-weather performance: Electric cars' batteries commonly lose charge in cold environments, further limiting their range.

They still pollute: If electric cars draw their power from power plants that are fueled by coal, their ultimate carbon footprint is comparable to that of the most efficient conventional cars.

They're not profitable: Car companies have struggled to make money from selling electric cars, in part because of consumer resistance but also because of high battery costs, the lack of a mature supply chain, and the fact that their billions of dollars of manufacturing capabilities are almost entirely oriented toward the production of cars based on a different propulsion technology—the internal combustion engine.

As you'll see over the course of this book, there are good answers to all of the questions raised by the points above, but forces in the automotive and oil industries have long dedicated themselves to making us think there can't be. Proponents of electric cars, until recently, had long been fighting a losing battle. Like them or not, gasoline vehicles were here to stay. Why bother attempting the impossible?

What those folks didn't know was that there would be a man who would make sport of creating companies that did what others said was impossible. They didn't know that someone could come along with enough money, enough intellect, and enough drive to upend everything the world thought it knew about electric cars. They didn't know about Elon Musk.

----~~~~~~----

As long as the laws of physics allow it, Musk believes it can be done. Before SpaceX, no private company had ever returned a spacecraft from low earth orbit. Before Tesla, few people believed it would be possible for a high-performance electric car to travel more than two hundred miles on a single charge of its battery. "One of Elon's greatest skills is the ability to pass off his vision as a mandate from heaven," Max Levchin, who cofounded PayPal with Musk, said in 2007. "He is very much the person who, when someone says it's impossible, shrugs and says, 'I think I can do it.'"

Musk spent the first seventeen years of his life in South Africa, growing up in the city of Pretoria. It was obvious from an early age that he was nerdy, reclusive, and determined. His parents sent him to school young, and, as the smallest child there, he attracted unwanted attention. Kids nicknamed him Muskrat. Turning inward, Musk often preferred the company of books to that of his peers, and lost himself in escapist sci-fi and fantasy, like Isaac Asimov's Foundation series and *The Lord of the Rings*. "The heroes of the books I read," he would say as an adult, "always felt a duty to save the world."

Musk's father, Errol, was an electrical and mechanical engineer who flew planes, sailed boats, and had an investment in an emerald mine in Zambia. His mother, Maye, was born in Canada to an American father and moved to South Africa with her family around 1950. She was, and continues to be, a model and nutritionist. Maye and Errol divorced when Musk was eight (Maye later characterized

it as her running away from Errol), and he spent three years moving from city to city with his mother and siblings. When he was
eleven, however, he decided to move back in with his dad in Pretoria.
Musk has said Errol wasn't a fun guy to be around—Musk's younger
sister, Tosca, called their father "very strict"—but it seemed like the
right thing to do because Errol had no kids at home. Many years
later, at sixty-eight, even Errol described himself as "an autocratic
father."

Errol dismissed computers as "toys that will amount to nothing,"
but Musk got his hands on one anyway and taught himself how to
code. At age twelve, he programmed a video game, which he called
Blastar, and sold the code to a computer magazine for $500. The
joystick-controlled game laid out its mission to players in clear terms:
"destroy alien freighter carrying deadly hydrogen bombs and status
beam machines." As a teenager, Musk continued to exhibit an entrepreneurial impulse, teaming up with his brother, Kimbal, fifteen
months his junior, to open a video arcade near his school. Despite
having a lease and suppliers all lined up, the brothers' plan was
thwarted when they found out they needed an adult's signature to get
the necessary business permit. Instead, the boys resigned themselves
to selling homemade chocolates to their classmates.

High school, however, was not fun. In those days, South Africa
could be a rough place to grow up, and Musk became a target of severe bullying. One incident hospitalized him for two weeks. He was
beaten so badly that his father didn't recognize him. "Kids gave Elon
a very hard time, and it had a huge impact on his life," Kimbal later
said.

All the while, Musk was looking to escape from apartheid South
Africa. He didn't want to be conscripted into compulsory service for
the South African Army—"Suppressing black people just didn't seem
like a really good way to spend time"—and he dreamed of life in
America, the center of innovation. "I would have come here from any

country," he said in 2007. "The US is where great things are possible." Ahead of his sixteenth birthday, Elon and Kimbal applied for Canadian passports, without bothering to inform their parents. Musk figured it was the easiest way to eventually get entry to the United States. In spite of his father's disapproval, he bought a flight to Canada the next year. He lived a frugal life there, subsisting on hot dogs and oranges, working odd jobs, and crashing at the homes of various relatives. In 1989, he found his way to Kingston, Ontario, and enrolled at Queen's University. His mother and siblings soon followed him to Canada. Errol remained in South Africa.

It was at Queen's that Musk met Justine Wilson, a budding writer from his Abnormal Psychology class who was, in Musk's words, an "intellectual with sort of an edge." He charmed her with chocolate chip ice cream, and they commenced an on-again, off-again romance. When Musk transferred to the Wharton business school at the University of Pennsylvania to study economics and physics, the two continued a long-distance relationship.

While at Wharton and finally starting to fulfill his American Dream, Musk wrote two papers that hinted at his future career. In one, titled "The Importance of Being Solar," he predicted the proliferation of solar technology. In the other, he dedicated forty-four pages to detailing how ultracapacitors could be used for energy storage, which would be helpful for things like electric cars.

Musk would exhibit this fascination with clean technology even in his personal life. In his 2015 biography of Musk, the journalist Ashlee Vance detailed an encounter Musk had with a young woman, Christie Nicholson, at his birthday party in Toronto. She was the daughter of a banking executive from whom Musk had sought business advice. He hadn't met her before his birthday. When Nicholson arrived at the party, Musk greeted her and led her to the couch. He didn't waste time with small talk. "I think a lot about electric cars," he said. "Do you think about electric cars?"

From its first day on the road, the Tesla Model S was a great electric car. But for Musk, who had made it his mission to replace every gasoline-burning vehicle on the road with an electric alternative, best in class simply wasn't good enough. To achieve Tesla's initial mission of accelerating the world's transition to sustainable transport, Musk's cars would have to be better than internal combustion engine vehicles in almost every regard.

At the time of my stomach-flattening drive through Napa with Dad, the Model S was a rear-wheel-drive car that performed well in snowy and icy conditions, but the market in parts of the world that experienced harsh winters still preferred all-wheel-drive vehicles. No all-wheel-drive electric car had ever been put into production. To Tesla, the challenge must have been too enticing to ignore.

A few months later, on October 9, 2014, Musk stood onstage at the Hawthorne Municipal Airport, next door to the Tesla Design Studio in a suburb of Los Angeles, and introduced the Model S P85D, a new mascot for the electric revolution. The car would have motors at the front and rear—a configuration that allowed it to distribute torque independently to each of the four tires. At the same time, its digital controls and highly responsive electric motors would give it precise regulation of traction in slippery conditions, with submillisecond re-action times. Not only would it be the fastest-accelerating sedan ever made—achieving a zero-to-sixty time of 3.2 seconds—but it would also enjoy traction control to match the best of the gasoline burners. The P85D further laid waste to the notion that the internal combustion engine should by default reign supreme.

"This car is nuts," Musk, in jeans and a dark dinner jacket, said onstage as thousands of Tesla owners and fans looked on from below. You could plant your foot on the accelerator and immediately get max power, he said. "It's like taking off from a carrier deck; it's just bananas." He fished for another analogy. "It's like having your own personal roller coaster that you can just use at any time."

Musk was in good spirits for what was arguably the most important night of Tesla's year. He had started his speech by making a joke that many people would consider unusual for a chief executive of a multibillion-dollar company. Eight days earlier, he had teased the launch of the P85D with a cryptic post on Twitter. "About time to unveil the D and something else," he tweeted. Within minutes, Twitter users and blog commenters were making fun of the tweet, choosing to interpret Musk's words in the most mischievous possible light. Onstage, Musk gamely acknowledged the flap. "There's been a lot of speculation about what the 'D' stands for," he said, waiting a beat before expanding his grin. "Yes, you'll notice my pants have Velcro seams." The crowd groaned and laughed at once.

Now he was reveling in the possibilities inherent in selling a car that behaved like a fighter jet. "Yeah, it's mad," he continued, with a dimpled grin. And then he added, "In the option selection, you'll be able to choose three settings: Normal, Sport, and Insane." A ripple of laughter washed over the crowd. Then, as if to reassure himself as much as everyone else: "It will actually say '*Insane.*'" He hunched his shoulders forward and laughed.

Videos posted by people who had experienced "Insane Mode" during test rides at the event appeared on YouTube the next day. Invariably, the accompanying commentary was littered with expletives and other delighted expressions of shock as the car's spine-straightening acceleration took effect. In the weeks and months that followed, more reaction videos appeared and spread, with one especially spicy compilation coming to accrue more than ten million views.

Insane Mode could be seen as more than just a product feature, more than just a marketing gimmick. It would be the mind-set required to fend off the short-sellers of Tesla's stock, traditional automakers, political opponents, and an increasingly nervous oil industry.

It represented the intensity of fervor needed to win the public over to electric cars. And it was a statement about the velocity of innovation required to transition the world to sustainable energy before the planet's climate changes beyond repair.

Even as a feature for a luxury motor vehicle, though, Insane Mode was audacious in both intent and implication. Few people in history could ever hope to pull it off. But Elon Musk had been burnishing his credentials for years.

―――〰〰〰―――

The distance between getting the crap kicked out of you as a South African schoolkid to being a billionaire space and auto industrialist in California is considerable, but Musk started to close the gap as soon as he moved to Silicon Valley. He got to Palo Alto, California, after leaving Wharton in 1995 (he ultimately graduated in 1997), intending to study at Stanford University toward a physics PhD with a focus on ultracapacitors. But once he saw what was going on around him, he had second thoughts.

At the time, start-up founders and venture capitalists were feverishly blowing the first breaths into the dot-com bubble. Companies with names like Netscape and Yahoo! were angling to be the next Microsoft or Oracle, and it became clear to Musk that the Internet was going to change the world.

Musk dropped his PhD plans and started his own Internet company, Zip2, which built an online mapping product that listed the details of businesses online—a prototypical Web version of the yellow pages. His brother, Kimbal, and a friend joined him later in the year. They rented a dingy office for $400 a month and lived and worked there, sleeping on futons and taking showers at the local YMCA. The Musk brothers poured every ounce of their energy into the company, but its investors insisted on installing an experienced professional as CEO. While Musk's share of the company was ultimately reduced to

7 percent after multiple financing rounds, he earned $22 million from the eventual sale of the company to Compaq for $307 million in February 1999—close to the peak of the bubble.

Musk celebrated the windfall by purchasing a million-dollar McLaren F1, one of the world's great supercars. A 1999 CNN documentary about Silicon Valley's newly minted millionaires followed Musk as the car was delivered to his house. "There are sixty-two McLarens in the world, and I will own one of them," the twenty-eight-year-old Musk intoned while dressed in an oversize mustard sport coat that gave him a distinctly dorky appearance (but hey—the nineties). "Just three years ago, I was showering at the Y and sleeping on the office floor. But now, obviously, I've got a million-dollar car and a few creature comforts." The suddenly wealthy wunderkind giggled with glee as Justine looked on. Sitting in the car with his girlfriend beside him, he said: "I'd say the real payoff is the sense of satisfaction of having created the company that I sold." And then Justine leaned in, put her arms around his neck, and said in his ear: "Yes, yes, yes, but the car is cool." Musk, nodding, added with a sheepish grin: "But the car sure is, sure is fun."

Musk's next company would deliver even greater returns. Seeking to create a full-service online bank, he cofounded X.com and invested $12 million of his own money in it. At first, the company limited its ambitions to focus on a feature that facilitated payments by e-mail, but it wasn't alone in this field. In 2000, after X.com got into a bidding war with rival Confinity to win customer market share on the online auction site eBay, the two start-ups decided to merge and become the market leader for e-mail-based payments. The resulting company was PayPal.

Musk's tenure as CEO of the merged entity was short-lived. After ten months in the job, he took a two-week trip to meet prospective investors and have a vacation in Sydney, Australia, with Justine, whom he had married in January 2000. While he was away, Confinity's founders, Peter Thiel and Max Levchin, staged a coup and convinced the board to remove Musk from the position. Officially, the falling-out centered on a disagreement about which software to use as the technology platform, but personality differences also came into play. Musk could be difficult to work with, Levchin said. "He's one of those guys who can be larger than the room." Thiel, who had left when X.com and Confinity merged, returned and assumed the CEO role.

Musk, however, remained PayPal's biggest shareholder, with an 11.7 percent stake in the company. When it eventually sold to eBay in 2002 for $1.5 billion, he earned a payout of about $180 million. It was with this money that he capitalized SpaceX and invested in an unknown electric car company called Tesla Motors.

3

THE FIGHT FOR THE ELECTRIC CAR

"If you're going through hell, keep going."

Even when he wasn't CEO of Tesla, Musk was a hands-on founder. He was instrumental in attracting other investors, influenced product design, and watched with dismay as the company struggled with cost overruns and quality issues in its early years. In August 2007, as chairman of Tesla's board of directors, it was Musk who delivered the news to founding CEO Martin Eberhard that he was being demoted. By December, Eberhard had left the company completely. Musk would ultimately take over the leadership duties himself, but not before looking for someone else to do it. By the end of 2007, Musk had interviewed at least twenty candidates for the role. He wanted a CEO to build Tesla into the next great automaker, but it was hard to find someone who understood start-ups and knew how to build hundreds of thousands of cars.

Reluctantly, and after two interim CEOs, Musk took over in October 2008, assuming the tripartite title CEO, Chairman, and Product

Architect (a mouthful he stuck with until 2014, when he became, simply, CEO). None of those positions would have seemed all that attractive in 2008. Immediately after taking leadership, Musk had to contend with the global financial crisis, all while trying to keep SpaceX, which had failed with its first three launch attempts, alive.

By that year, he and Justine had five sons—a pair of twins and a set of triplets—but were experiencing marital difficulties. Musk filed for divorce and within weeks started a relationship with the young British actress Talulah Riley. Captivated by Riley, Musk quickly proposed marriage and was accepted. (They married in 2010.)* By this point, Musk was living in Los Angeles and had undergone a transformation.

The figure Musk cut was no longer that of the stereotypical, fashion-averse software engineer but one fit for the covers of national magazines and appearances on late-night talk shows. Soon enough, he was featuring on both. In December 2008, he was the subject of an admiring profile in *GQ* that was headlined "The Believer." The accompanying illustration showed Musk's head above the clouds—not *in* them, note—and looking out to space. In April 2009, Musk was given a guest slot on the *Late Show with David Letterman* to discuss the Model S concept car. The same year, he was the subject of a *New Yorker* profile for which he was photographed with his five sons in front of a clay model of the car.

More magazine profiles were to come—*Wired* (2010), *Forbes* (2012), *Esquire* (2012), *Fortune* (2013)—and Musk was a central figure in the 2011 documentary *Revenge of the Electric Car*, which showed Tesla as it struggled through the financial crisis, marketed the Roadster, and lifted the cover on the Model S. Behind the glamorous cover stories and TV appearances, however, was a grit that the words and pictures

---

* The couple ultimately divorced (for the second time) in 2016.

couldn't quite capture. For Musk to get his companies this far, he had to scrap, claw, and hustle.

It is no secret that Musk is an unconventional leader. He has been described by current and former employees as simultaneously daring, delightful, and difficult. "Elon always wants to know, 'Why are we not going faster?'" one of his employees told writer Tim Urban. "He always wants bigger, better, faster." Even JB Straubel, Musk's cofounder and Tesla's chief technology officer, has called the boss "an interesting mix of extremely challenging and incredibly difficult." But perhaps Musk's most important trait as a CEO has been his ability to overcome tough times. "If you're going through hell," he has said, citing one of his favorite quotations, "keep going."

These qualities have been essential in getting Tesla where it is today. Without them, in fact, Tesla probably wouldn't have outlived the Roadster era. Building a new business is hard in any context. "Starting a company," Musk has said, paraphrasing his friend Bill Lee, an entrepreneur and investor, "is like eating glass and staring into the abyss." But it's even harder if you're trying to enter the auto industry, where the incumbents have been protected by high barriers to entry, such as the costs of building manufacturing plants, finding quality suppliers willing to work with small production numbers, and establishing distribution networks. Starting an *electric* car company, meanwhile, is a challenge on an altogether different level. History has been no friend to electric cars. Not even Thomas Edison, working in more favorable conditions, could make them succeed.

"Electricity is the thing," Edison said in 1903.

> There are no whirring and grinding gears with their numerous levers to confuse. There is not that almost terrifying uncertain throb and whirr of the powerful combustion engine. There is no water-circulating system to get

out of order—no dangerous and evil-smelling gasoline and
no noise.

As the automotive era was getting started, Edison had decided that
electric cars were the way of the future. Had he known what was
ahead, he might have been surprised by how long the future would
take to arrive. He did what he could to help the technology along, in
no small part because he had developed nickel-iron batteries that
would serve just the purpose. In 1901, he claimed he had developed an
electric car that could reach speeds of seventy miles per hour. The following year, he said the prototype could drive eighty-five miles on a
single charge and vowed to get his battery to market within a matter
of months. "It will be but a short time before demand for storage batteries will create one of the most enormous industries in the land," he
said at the time. Musk would make similar claims 113 years later.

Ultimately, Edison couldn't deliver on his hype. The groundbreaking electric car he promised never went into production. But he wasn't
deterred. In 1914, he teamed up with his friend Henry Ford in another
effort to make a battery that would be suitable for an electric car.

"Within a year, I hope, we shall begin the manufacture of an electric automobile," Ford told *The New York Times* in January 1914. He
didn't want to reveal too much of his plans, but Ford confirmed rumors of his partnership with Edison.

> The fact is that Mr. Edison and I have been working for
> some years on an electric automobile which would be
> cheap and practicable. Cars have been built for experimen-
> tal purposes, and we are satisfied now that the way is clear
> to success. The problem so far has been to build a storage
> battery of light weight which would operate for long dis-
> tances without recharging. Mr. Edison has been experi-
> menting with such a battery for some time.

It's unclear if there really were multiple prototypes at the time, but there was at least one. A photo from 1913 shows it parked outside Ford's Highland Park plant in Michigan. In it, an electrical engineer from Ford sits in a carriage seat perched atop a box that encloses three suitcase-size nickel-iron batteries. The driver's seat is built upon a simple frame that ramps up to the axles at either end of the vehicle. The steering would be handled as it is in a small boat with a tiller. A photo from 1914 shows a second electric car, this one built on a Model T frame with a Model T steering wheel, and the batteries again under the driver's seat. Rumor had it that the car, due to hit the road by 1916, would be priced between $500 and $750 (between $11,000 and $16,000 today) and would have a range of up to a hundred miles per charge.

Edison and Ford's creation was not an anomaly. At the dawn of the automotive era, it wasn't clear that gasoline-powered cars would prevail over their steam-powered and electric counterparts. Indeed, for a while, electric cars held the advantage. Steam-powered cars, which had been around since the 1700s, were the early incumbents, but they couldn't travel far without needing to be refilled with water, and they could take up to forty-five minutes to start. The first internal combustion engine cars were designed in the early 1800s, but they were inconvenient to drive. They had to be started with hand cranks, and their gears needed to be changed manually. They were also noisy and dirty, emitting an exhaust of nitrogen oxides, carbon dioxide, and soot wherever they went.

Electric cars, on the other hand, were exhaust-free, easy to drive, and quiet. Some of the leading automotive innovators poured their energies into making them work. It was a Scotsman, Robert Anderson, who came up with the first electric carriages in the 1830s. It took more than fifty years, though, for the technology to reach the United States, and that was thanks to another Scot, a chemist named William Morrison. Morrison unveiled his four-horsepower "horseless carriage"

in Des Moines, Iowa, in the late 1880s. In the 1890s, Connecticut's
Pope Manufacturing Company became the first electric vehicle manufacturer, and a young electrical engineer named Ferdinand Porsche
designed an electric vehicle for the Austrian carriage maker Ludwig
Lohner. Porsche, who would later start a sports car company that bore
his name, put the so-called P1 on the streets of Vienna in 1898. The P1
boasted speeds of twenty-one miles an hour and a range of forty-nine
miles per charge.

Despite the hype and their combined star power, the "cheap and
practicable" electric car that Ford and Edison promised never eventu-
ated. Edison couldn't get his batteries to pass the testing phase, and
he failed even to develop one that could power an internal combustion
engine car's starter motor. Ford, who had other things to deal with,
eventually gave up on the plan to work with his friend, despite initial
plans to buy a hundred thousand batteries from the inventor and invest $1.5 million ($36 million today) in the project.

Hopes for electric cars were dealt a killer blow when, just as Edison
failed to make his batteries work, Charles Kettering perfected the
design on an electric starter motor that eliminated the need to start a
gasoline car by hand crank. Suddenly, internal combustion engine
vehicles seemed eminently more practical, not least because of a lack
of electrical infrastructure in the United States. Ford turned his full
attention to mass manufacturing the gasoline burners, and cheap oil
from the Texas oil fields, discovered in the early 1900s, fueled a momentum shift that made electric cars essentially obsolete by the 1930s.

Anyone who still believes in the need to fight for electric cars has
fair reason. Six decades after they disappeared from America's roads
the first time, they made a sudden resurgence—only to die again.

The revived hopes for electric cars started in California. In an attempt to curb air pollution, the California Air Resources Board
(CARB) established a rule that required automakers who sold cars in
the state to offer zero-emissions vehicles. CARB was inspired in part

by GM, which had produced an all-electric concept car called the Impact. In 1996, GM started leasing a production version of that car in California and Arizona under the name EV1.

The EV1 was a sporty two-door model that could accelerate from zero to sixty miles per hour in 6.3 seconds. Drivers gained entry to the car with a keypad instead of a key. It quickly found a cult following among clean-tech enthusiasts and was driven by film stars such as Tom Hanks and Mel Gibson. However, no one could own the vehicle. Just 1,117 EV1s were ever made, and they were offered by lease only, ostensibly so GM could ensure quality repairs. The automaker soon found reason to end even that meager offering. In 1999, citing lack of public demand and high production costs, GM stopped manufacturing the EV1. Three years later, it discontinued the lease program and sent all but a few cars to the demolition yard. The rest were destined for museums and universities.

The real reasons for the EV1's demise were likely more complex. The 2006 documentary *Who Killed the Electric Car?* pinned the blame not only on consumers and automakers but also on President George W. Bush's administration and the oil industry. While consumers were starting a love affair with SUVs and showed little interest in range-limited EVs, automakers focused their resources on more-profitable gas guzzlers. In 2002, for instance, GM was selling nearly four thousand Hummers a month. Environmental credentials weren't the Hummer's strong point—it eked out just fourteen miles per gallon.

Meanwhile, at the time of the EV1's death, the Bush administration pushed CARB to abandon its electric car mandate. In 2002, the administration's Department of Justice supported a lawsuit brought by GM and DaimlerChrysler against CARB that claimed California was attempting to set a mileage standard. Such a standard was (and continues to be) the domain of the federal government, the parties argued. President Bush's chief of staff at the time, Andrew Card, had been chief lobbyist for GM and was head of an auto industry trade association when California first proposed the mandate (which he opposed). White House spokesman Scott McClellan denied that the administration was taking sides because of Card's connections to the auto industry. "The American people would be best served if the leadership of special interest groups worked with us in our efforts to increase fuel efficiency, promote safety, and improve air quality," McClellan said in October 2002. In 2003, CARB relaxed its emissions standards and dropped the electric car requirement.

Oil companies were in on the action, too. The Western States Petroleum Association, an oil industry lobby group, had funded a campaign against utilities that were installing electric car chargers. Through proxies called Californians Against Utility Company Abuse and the Clean Air Alliance, the lobby group urged a ratepayer revolt by calling the planned charging infrastructure a "hidden tax" on electric bills. It also disseminated technical and economic arguments against electric cars.

When GM's EVis were ultimately crushed, so were the hopes of electric car advocates. A band of supporters turned up to the wrecking yard and staged a candlelit memorial for their lost vehicular friends. Elsewhere, an interested observer was taking note. "When was the last time you heard about any company's customers holding a candlelit vigil for the demise of their product?" Elon Musk asked in a January 2013 interview. "Particularly a GM product!" The saga, he said, served as inspiration for his investment in Tesla.

Musk has characterized his companies as moral missions as much as businesses. He didn't start Tesla or SpaceX to make money, he has said, but because he believed the world needed them. The future for humans on Earth would be terrible if we didn't switch to sustainable energy, and without electric cars, the peril from climate change would be unimaginable. His goal to colonize Mars is also motivated in part by a moral impulse. In the case of an extinction event, which could be brought on by anything from runaway climate change to rogue artificial intelligence, we'd all be homeless. "I think there is a strong humanitarian argument for making life multiplanetary," he has said, "in order to safeguard the existence of humanity in the event that something catastrophic were to happen."

The degree to which he cares about his companies on a deep, personal level was brought home in a 2014 interview in a Tesla store in London. Asked by a reporter about his sensitivity to criticism of his companies, Musk compared the experience to a parent having his child unfairly maligned. "There are honest criticisms to be had, certainly," he said. "But it's difficult to take false criticism of something you care about."

Guided by this view of his work and determined to protect the future of electric cars, Musk has made a habit of responding to perceived slights. Let's take a quick look at the Elon Musk Pugilistic Powerwall of Fame.

Henrik Fisker: Tesla sued the Danish designer after hiring him on contract to do the styling for an electric sedan, code-named White Star (the program would later spawn the Model S). An arbitrator found in Fisker's favor and ordered Tesla to pay more than $1 million in legal fees. "I don't think very highly of Henrik Fisker," Musk would say in a 2012 interview. At the same time, he called the Fisker Karma, a luxury hybrid sports car, "a mediocre product at a high price."

Martin Eberhard: Eberhard, Tesla's first CEO and one of its founders, sued his former company for breach of contract and slander. Musk responded by calling Eberhard a liar and published a blog post detailing a list of decisions made by the former CEO that sent the Roadster project way over budget. The parties settled the suit out of court and didn't disclose the terms of the settlement.

Top Gear: The BBC car show presented an unflattering image of the Tesla Roadster, showing the hosts pushing it by hand after it had supposedly run out of charge. Musk called the show "about as authentic as a Milli Vanilli concert," and Tesla sued *Top Gear* for libel. After losing in court twice, Tesla appealed. A judge dismissed that, too.

John Broder: Musk called Broder's *New York Times* story, about stalling on the side of a highway in a Model S, "fake" and published a blog post saying that some journalists believe "facts shouldn't get in the way of a salacious story."

Randall Stross: After the writer published a column in *The New York Times* that questioned why taxpayers were subsidizing a company that sold expensive cars, Musk called Stross a "huge douchebag" and an "idiot."

George Clooney: The Hollywood actor griped to *Esquire* that his Tesla Roadster kept getting stuck on the side of the road. In response, Musk tweeted: "In other news, George Clooney reports that his iPhone 1 had a bug back in '07."

Jarrett Walker: Public transport advocate Jarrett Walker took issue with Musk's saying in an interview that public transport sucked. Walker tweeted that Musk's apparent disdain for public transport was a "luxury (or pathology) that only the rich can afford." Musk responded on Twitter by calling Walker an "idiot." But then he apologized and corrected himself, clarifying that he meant to say "sanctimonious idiot."

Mitt Romney: In a debate with President Obama, the Republican Party's 2012 presidential nominee named Tesla among a group of clean-tech companies that had received government funding, labeling them "losers." Romney "was right about the object of that statement," Musk would later say, "but not the subject."

Since his teenage years, Musk has known what it's like to be the victim of bullies. His responses to critics, fair or not, suggest he has no intention of ever again putting himself in such a position of submission. At the same time, he also seems to have paid heed to the lessons of history.

Some of Musk's critics have called him a modern-day Preston Tucker, the entrepreneur-engineer who, in the 1940s, started a company to build a car that would introduce innovations such as safety belts, windows that popped out on impact, and an engine in the back for rear-wheel drive. Tucker, part hype artist and part genius, produced fifty prototypes of his glamorous Tucker sedan but was charged with fraud by the Securities and Exchange Commission (SEC), which, according to a report leaked to the media, questioned his intention of mass manufacturing the vehicle. While Tucker ultimately won in court, his company, buried with debt and facing lawsuits from dealers who had been promised delivery of the new sedan, could not withstand the pressure. It closed down as the trial began. Tucker died from lung cancer in 1956, aged fifty-three.

A popular conspiracy theory holds that Tucker was the victim of the Big Three—GM, Ford, and Chrysler—working with their associates in the SEC to neuter a threat to their businesses. Whether or not there is any truth to that assertion, six decades later, the institutional opposition to Tesla's efforts to change the automotive industry has been overt.

Throughout its history, Tesla has had to contend with a continuation of the obstacles that thwarted electric cars over the past 150 years. However, it has also faced a series of fresh challenges. As Tesla moved beyond the Roadster and closer to becoming a mainstream automaker with the Model S, it would find itself fighting to protect its safety record in the wake of a series of high-profile fires; to defend its sales model against auto dealers; and to reassure skeptical consumers about the long-distance drivability of its cars in the face of entrenched "range anxiety."

Each new battle would test the depths of the CEO's force of will. Edison couldn't do it, but maybe Musk could.

――――~~~~~~~――――

As a business leader, Musk shares at least one thing in common with Steve Jobs: He is a wartime CEO. Such leaders preside over their companies when they face an imminent existential threat, says Ben Horowitz, a Silicon Valley investor and partner with Netscape cofounder Marc Andreessen in the celebrated venture capital firm Andreessen Horowitz. The wartime CEO's companies depend on strict adherence to a mission, Horowitz wrote in his book, *The Hard Thing about Hard Things*. While a peacetime CEO follows protocol, a wartime CEO violates it in order to win. While a peacetime CEO defines company culture, the wartime CEO lets the war define the culture. A peacetime CEO knows what to do with a big advantage, while a wartime CEO is paranoid.

"We had multiple near-death experiences," Musk has said of Tesla's history. "Death on the nose—just right in front of you."

Peacetime leaders must broaden the current business opportunity, so they encourage creativity and contribution across a diverse set of objectives, wrote Horowitz. "In wartime, by contrast, the company typically has a single bullet in the chamber and must, at all costs, hit the target."

"If you are fighting a battle, it's way better if you are at the front lines," Musk said, explaining why he had a desk on the factory floor when the Model S started production. "A general behind the lines is going to lose."

"Wartime CEO," wrote Horowitz, "cares about a speck of dust on a gnat's ass if it interferes with the prime directive."

"If we had not succeeded, then we would have been used as a counter-example of why people shouldn't do electric cars," Musk said about Tesla's near-bankruptcy during the financial crisis. "People would have used Tesla as an example of just another stupid car company, basically."

Horowitz was echoing the former Intel CEO Andy Grove, a business hero in Silicon Valley, who had written: "Success breeds complacency. Complacency breeds failure. Only the paranoid survive." The threats to a company can come from competition, dramatic macroeconomic change, market change, supply chain change, and so on, Horowitz said. Tesla has faced it all.

"When Henry Ford made cheap, reliable cars, people said, 'Nah, what's wrong with a horse?' That was a huge bet he made, and it worked," Musk said in 2013.

Management books tend to be written by management consultants who study companies during their times of peace, Horowitz cautioned. Other than Grove's writing, he didn't know of any books that taught people how to manage in wartime like Steve Jobs.

When Jobs returned to Apple as interim CEO in 1997, the company was months away from bankruptcy. Four years later, it released the iPod.

On May 8, 2013, Tesla reported its first quarterly profit. There had been many start-ups over the decades, but, at last, an electric car company was selling enough cars to stay in business. Profitability was what made a company real, Musk said. "Tesla is here to stay and keep fighting for the electric car revolution."

But that fight was just getting started.

4

ON FIRE

"What the heck is going on?"

Lithium-ion batteries are wonderful things. As well as being light-weight, they run a long time between charges, recharge quickly, can be charged and discharged many times without suffering significant capacity loss, and they're among the highest-density batteries the world knows today. They fit neatly in your laptop or mobile phone, and they're a key reason Tesla even exists.

But they are also highly flammable, and when they catch fire, there's nothing subtle about the explosion. A blazing lithium-ion battery sends plump, pillowy flames of brilliant colors and plumes of black smoke high into the air. Upon ignition, the batteries can also act like erratic fireworks, shooting out sparks and fireballs without warning.

In 2006, a few fires in laptops that used Sony-made batteries, hundreds of which overheated, led to a recall of millions of lithium-ion battery packs. In 2013, a mechanic found flames and smoke spewing

out from the auxiliary power unit of a Boeing 787 Dreamliner parked
at Boston's Logan International Airport. The culprit was thermal runaway in lithium-ion batteries. After another 787 experienced a battery-
malfunction warning while in the air five days later, Boeing was
forced to ground its entire Dreamliner fleet to fix the problem. In
2015, airlines banned so-called hoverboards—battery-powered ride-on
vehicles designed for short-distance solo travel—because of their pro-
pensity to overheat and combust. The next year, the New York Police
Department tweeted photos of pillows that had been set alight by
spontaneously combusting smartphones, and Samsung was forced to
stop selling its Galaxy Note 7 after several battery-related fires.

Each lithium-ion battery cell contains a highly flammable liquid
electrolyte (a chemical medium that facilitates the flow of electrical
charge) that can ignite if a battery's electrodes get too hot—as can
happen if they short-circuit when damaged.

When the batteries are used properly, such fires are rare, occurring
in one in a hundred million cells. The fire risk can also be reduced by
cooling systems. Tesla's engineers developed a system to keep the batteries cool with liquid glycol—a kind of antifreeze commonly used to
keep car engines from freezing in winter and overheating in summer—
which snakes its way through the battery pack in a metallic tube.
The system is designed to quickly chill the cells and separate them
from one another, so that if one catches fire, its neighbors remain un-
affected.

But as with any system, things can still go wrong.

In October 2013, things started to go wrong very quickly. Three
Model S fires in the space of six weeks brought Tesla's run of luck to
a sudden stop and cast a shadow on its reputation. The fires would be
the most explosive of three key threats that Tesla would have to overcome in its adolescence to cement its legitimacy and win more widespread public support. Fights with auto dealers over the right to sell
cars directly to buyers would consume much of Tesla's legal energy in

the midterm, and over the long term, it would have to prove naysayers wrong by showing that its cars could easily handle long-distance travel. But in late 2013, the fire fight was its most urgent crisis.

The first of the combustible events happened in the first week of October, when a Model S struck a metallic object on a highway near Seattle, Washington. The car's battery pack was punctured when the object was squeezed vertically between the road and the underside of the car. Once the car's computer system detected that some of the battery modules had been damaged, it told the driver to pull over. A fire started in the modules after the driver had exited the vehicle. While the fire was contained to the front of the car—a safety measure implemented by Tesla's engineers—firefighters inadvertently encouraged the blaze by cutting into the battery pack's metal firewall and spraying it with water, creating more holes through which the flames could vent. The result was a spectacular inferno that looked alarming when it was inevitably caught on camera by a passing motorist, who exclaimed, "Oh, that's a Tesla, dude!" The video was featured on the TV news, sending investors into a panic. After a months-long surge that had taken Tesla's share price from $29 to $190, the stock slid by 10 percent.

Next was another photogenic blaze three weeks later in Mérida, Mexico, resulting from a dramatic high-speed crash. The car had hit 110 miles per hour before plowing into a roundabout, shearing off fifteen feet of curb, bulldozing a concrete wall, and smashing into a tree. The driver escaped injury, even as the car lost two of its wheels. Again, the fire, which started several minutes after the driver left the vehicle, was contained to a small area in the car's front. Tesla wasn't blamed for the fire because the wreck was so extreme, but its share price dropped by 3 percent anyway.

When another fire occurred just three weeks later, Tesla really started to feel the heat. This time, a driver in Tennessee drove over a three-ball tow hitch at seventy miles an hour on the highway. The tow

hitch wedged itself between the road and the car's belly, then punctured the battery pack. The computer system warned the driver to pull
over, and seconds after he exited, smoke started pouring out. Flames
followed minutes later. Again, however, the fire was restricted to the
front of the vehicle, and there was no damage to the cabin.

The media impact was instant. Three months earlier, Tesla had
issued a press release celebrating a five-star safety rating from the National Highway Traffic Safety Administration, noting that it was the
highest rating ever given. After heralding that achievement, though,
the media were now looking at Tesla through a more skeptical lens.
Reporters wondered if NHTSA would issue a Model S recall and
speculated that Tesla's reputation could be irreparably harmed. The
fires, according to *ABC News*, raised doubts about a "seemingly un-
touchable company."

Investors followed the media's cues. In the week after the Tennessee fire, Tesla's stock price tumbled another $15 per share. Over the
course of two months, the company's market valuation had been
slashed by $8 billion, the share price shrinking from a high of $191 to
$121. NHTSA opened an investigation into the matter, shining a spot-
light on the safety of the Model S. Appearing in an interview on
CNN in the midst of the drama, Musk said he felt like he had been
"pistol-whipped." "Our three noninjurious fires got more national
headlines than a quarter-million deadly gasoline-car fires," he said,
holding his arms out wide. "That's mad! What the heck is going on?"

While the media ran with the story, speculating on the existential
threat the fires posed to electric cars, Musk turned to the Tesla blog.
He wrote a post that contextualized the fires and laid out the company's mission—to accelerate the world's transition to sustainable
transport—emphasizing what was at stake if the public were to turn
against electric cars.

Musk pointed out that there had been four hundred deaths from
more than 250,000 gasoline-car fires in the United States alone in the

four years since the Model S went into production. In the same period, there had been no deaths or serious injuries in a Model S. "The
media coverage of Model S fires vs. gasoline car fires is disproportion-
ate by several orders of magnitude," he wrote, "despite the latter actually being far more deadly." To reduce the chances of fires from
battery-pack punctures, Tesla would update software in the Model S
to raise ground clearance when driving at high speeds. At the same
time, the company started designing shields that would sit on the
underside of the cars to provide further protection for the battery
packs.

The company finished production of the shields not long before
NHTSA was due to deliver its verdicts on Tesla's culpability for the
fires in Washington and Tennessee. The shields were designed to de-
flect and crush road debris that passed under the vehicle. In tests,
Tesla's cameras showed the shields, which were made up of aluminum
bars and a titanium plate, smashing tow-bar hitches, alternators, and
concrete blocks into tiny pieces.

To introduce the shields to the world and emphasize the Model S's
resistance to fire, Musk again wrote a blog post. This time, he relied
on more than just strong language to make his point. He included
slow-motion videos of the shields obliterating the foreign objects. "We
believe these changes will also help prevent a fire resulting from an
extremely high speed impact that tears the wheels off the car," like the
one in Mexico, Musk wrote. The same day, NHTSA announced that
it was closing its investigation into the fires and reported that it had
not identified a safety defect trend that would justify a recall.

As with other cars, there would occasionally be more Tesla fires. In
February 2014, a parked Model S combusted, burning down itself and
the garage it was in. In January 2016, a Model S burned down while at
a Supercharger in Norway. In August 2016, a Model S caught fire
during a test-drive event in France. In November 2016, a Model S in
Indianapolis smashed into a tree at high speed and was engulfed in

flames, killing its two occupants. In March 2017, a parked Model S caught fire in Yorkshire, England; and in October 2017, a Model S went up in flames after a highway smash, attracting thirty-five firefighters to the scene.

Despite the vivid images of violent blazes and grotesquely melted cars, however, the stock market and the media, willing to accept the logic that other cars catch fire, too, tempered their previous panic. Tesla's business continued as normal. Fire, it seemed, was no longer a threat to the company.

5

JUST DEAL

"Tesla finishes last in being salesy! Good."

As often seems to be the case with Tesla, just when one problem was resolved, it was promptly replaced by another. As 2014 rolled on, Tesla faced intensifying questions about its right to sell its own products.

In 2007, Musk envisaged Tesla's stores as combining the best of Starbucks, Apple stores, and "a good restaurant," writing in a blog post that the company planned to put as much energy into making the stores look good as it did with its cars. Tesla stores today are sleek and modern, outfitted with interactive touch screens, Keurig coffee machines, and LCD monitors that show its cars gliding through sunsets.

Still, some states think they are illegal.

Automobile dealers associations have objected to Tesla's direct-sales strategy, claiming that Musk's company is breaking the law by cutting out the middleman. In some states, including Texas, Michigan, and

Connecticut, long-standing franchise laws grant dealers the exclusive right to sell new cars.

The dealer system stands in the way of Tesla's business on several fronts. The company has said the primary reason it sells directly to consumers is that it wants to control the way its products are brought to market. That's particularly important when it comes to selling electric cars because most people don't know as much about them as they do about gasoline cars. The company has called its stores "education venues" as much as retail venues. Tesla is also intent on controlling its brand. Apple, for example, has been careful in establishing its own retail network and for a long time didn't allow any other stores to sell iPhones. Tesla has taken a similar path.

The luxe act of perusing a Tesla store, where the salespeople are told not to pressure customers into a purchase, contrasts with the experience of visiting a conventional car dealership, where price negotiations and overattentive service are common. A 2015 survey by online retailer Autotrader (admittedly, a biased source) found that just 17 of 4,002 respondents liked the car-buying process as it was. In July 2016, after a mystery shopper survey found that Tesla ranked last among auto brands for dealership sales effectiveness, Musk celebrated with a tweet: "Tesla finishes last in being salesy! Good."

Multiple surveys have suggested that dealers generally aren't skilled at selling electric vehicles—perhaps because many of them don't want to be. In late 2013 and early 2014, *Consumer Reports* conducted a secret survey of eighty-five dealerships in four states and found that "many dealership salespeople were not as knowledgeable about electric cars as you might expect." Thirteen of the eighty-five dealers actively discouraged the sale of electric cars, and thirty-five recommended buying a gasoline car instead. A 2016 mystery shopper survey of 308 dealerships by volunteers from the Sierra Club found that many salespeople didn't know about tax credits and rebates for electric cars, nor how much it cost to operate them. My wife had this experience on a call with a Nissan dealer in 2017, when she was trying to find out what rebates she'd qualify for if she bought the Leaf. The respondent had no idea.

In 2015, the former chairman of the National Automobile Dealers Association compared electric cars to broccoli, when consumers really want "low-calorie doughnuts"—fuel-efficient gas cars. A closer look at how dealers operate, however, suggests that consumers aren't entirely to blame for their supposed preference for the vehicular equivalent of fried batter. While the public is still getting used to the concept of a car that doesn't run on gasoline, it takes dealers longer to explain what electric cars are and how they work. A Nissan spokesman told *The New York Times* that a salesperson "can sell two gas burners in less than it takes to sell a Leaf." Meanwhile, electric cars, with fewer moving parts and theoretically less need for maintenance, also pose a threat to the dealers' service departments, their number one source of profits. It's hardly surprising, then, that Tesla would want to take matters into its own hands.

The origins of Tesla's battle with the dealers can be traced back to an economic recession in 1920 and the Great Depression in the 1930s. In those times, dealers were at the mercy of automakers, particularly Ford and General Motors, who offloaded surplus stock on their franchisees even though demand was extremely lean.

The dealer model helped accelerate car sales by allowing manufacturers an easy way to reach almost every customer in the country through incentivized third parties, thereby attaining maximum regional spread. But the automakers would ultimately be stripped of their right to sell directly to consumers. In 1937, states began enacting laws that would prevent manufacturers from exploiting dealers as they did during the Depression. Dealers now rely on those laws to fight Tesla.

In the United States, dealers have political muscle. New-car dealerships account for 15 percent of all US sales tax revenue, according to the National Automobile Dealers Association—and in some states, that figure is as high as 20 percent. In almost every town in America, too, they're among the most prominent symbols of economic power. Collectively, dealerships employ more than a million Americans, and they spend big on advertising, events, and groups in their communities, including Little League teams. They also donate to political campaigns, even at the national level. In the 2012 election cycle, according to OpenSecrets.org, the National Automobile Dealers Association spent more than $3.2 million on campaign contributions. The same year, the group spent $3.49 million on lobbying efforts. At the local level, dealers can be found in almost every congressional district, where the effects of their economic contribution are more keenly felt.

Essentially since auto sales began, dealers have served as chaperones helping millions of Americans make one of the most important purchase decisions of their lives, perhaps second only to buying a house. It is difficult to overstate the significance of the private car in the American psyche. For many, a car is not only a practical means of transportation but also a symbol of status, freedom, and independence. The country's fortunes rose commensurately with the proliferation of the automobile, as did a sense of national and personal identity. You might not own a house, but you can own a car. If the dealers were to cede their protected status as exclusive retailers of automobiles, they would undermine their business model as well as their privileged position in American culture.

These are the reasons why Tesla has faced difficulty in pursuing its direct-sales model.

The dealers associations argue that they protect consumers by preserving price competition among dealerships and offering more service options when cars break down. The US Federal Trade Commission has disagreed with that assessment. In a blog post published in 2015, its staffers wrote: "A fundamental principle of competition is that consumers—not regulation—should determine what they buy and how they buy it." Nevertheless, the dealers' position has some powerful support. When, in 2014, Governor Rick Snyder signed a bill that blocked Tesla from selling cars directly to consumers in Michigan, GM issued a statement applauding the action. The law, GM said, "will ensure we compete under the same rules in the marketplace as other automobile manufacturers."

Tesla doesn't think the franchise laws are relevant, because they concern only franchisor-franchisee relationships. While the laws prevent, say, General Motors from selling cars at its own stores and thus cannibalizing its dealerships, Tesla has no interest in starting a franchise. It just wants to sell cars itself.

Tesla's position received a boost in September 2014, when the Massachusetts Supreme Judicial Court agreed with the company in a ruling that blocked the Massachusetts State Automobile Dealers Association's attempt to shut down a Tesla store near Boston. In her decision, Justice Margot Botsford wrote that the law was intended to protect dealers only from unfair practices of manufacturers and distributors "with which they are associated, generally in a franchise relationship," and not unaffiliated manufacturers. Tesla was quick to praise the decision and tie it to disputes in other states. "We have battles in New Jersey and other states with similar constructs, and we hope and expect the same interpretation would carry over to those venues," Tesla's deputy general counsel said at the time.

The battle in New Jersey was to provide a high-profile example of how Tesla intended to deal with the ongoing challenge. In March 2014, the state government used an unusual tactic to bar Tesla from selling cars at its stores there. Without full public notice, the New Jersey Motor Vehicle Commission—of which half of the eight members were Governor Chris Christie's political appointees—had voted to rescind two sales licenses the state had previously granted to Tesla. Dozens of Tesla's supporters went to the meeting to protest the move but weren't allowed to speak until the vote had already gone through.

After the ruling, Tesla had to change its existing stores in the state to "galleries," where it couldn't offer test drives or discuss any pricing information. Customers instead had to order their cars online and get them shipped in from out of state. This is also what Tesla has had to do in other states where it has been barred from direct sales.

A few days later, Musk took up the cause by writing a post that was addressed to "the people of New Jersey." He took square aim at Christie, who had been implicated in a scandal in which his staff arranged for restricted access to the George Washington Bridge from New Jersey to Manhattan as an act of political retribution against a local mayor, whose constituents were then locked in traffic jams for days.

> The rationale given for the regulation change that requires auto companies to sell through dealers is that it ensures "consumer protection." If you believe this, Gov. Christie has a bridge closure he wants to sell you! Unless they are referring to the mafia version of "protection," this is obviously untrue.

The "mafia" line was bold, given the special significance of the term in the state Christie governed and in which the TV series *The Sopranos* was set. But it paid off. The ensuing headlines marveled at Musk's temerity. "Oh No He Didn't," wrote the *Wall Street Journal*'s *MoneyBeat* blog. "Elon Musk Throws Mafia, Bridge References at Gov. Christie." In one paragraph, Musk had turned a local fight into a national cause and positioned Tesla as a plucky underdog just trying to get by in a system that had been rigged to work against it.

By March 2015, Governor Christie had signed a bill to reverse the ban.

6

RANGE ANXIETY

"Travel for free, forever, on pure sunlight!"

When I drove my dad around Napa Valley in a Model S, I found it difficult to relax. Throughout the trip, despite views of rolling hills and expansive vineyards, I kept watching the car's state of charge indicated on the small touch screen, behind the steering wheel, that served as a digital instrument cluster. Below the speedometer in the center of the screen, the vehicle told me its "rated range," which was calculated according to the US Environmental Protection Agency's testing of how far the car could travel on each charge of its battery in standard driving conditions. The EPA had found that the Model S had a range of 265 miles per charge. However, if you drove aggressively, or went up a lot of hills, or faced strong headwinds, you could expect the car to deplete its charge well short of the 265-mile mark.

Conversely, if you drove conservatively, went down a lot of hills, or benefited from tailwinds, you could expect to travel a good deal farther than 265 miles. Driving a Tesla is like an energy-trading game—you mostly use up energy from the battery in the name of acceleration, but you can also return energy to the battery through its regenerative braking system, which kicks in any time you take your foot off the accelerator, an action that slows the car and converts its kinetic energy into chemical energy. This process helps extend the car's range.

Even though the car was telling me that we had 120 miles of rated range remaining, I figured that, with the sixty miles back to San Francisco, and then the thirty miles I would need to drive to work the next day, I would have to play it safe. That thirty-mile range buffer could be wiped out by one too many stomps on the accelerator.

When I'm anxious, I do this little neck-stretching thing, or I use the thumbnail on each hand to systematically apply pressure to each one of my fingertips in an allegro tempo, usually on repeat, like a loop of low-burn hyperactivity. As I was driving that day, I was doing both those things.

I was nervous because the prospect of being stranded on the side of the road with zero energy was unappealing. In such an event, we'd have no choice but to have the Model S towed or to push it by hand in a heads-down walk of shame to the nearest power outlet. We might have been able to find an outlet at a gas station or in a friendly stranger's garage, but those would have been undesirable outcomes. Aside from the awkwardness of having to explain to a stranger why our $100,000 car was undrivable, we'd have to twiddle our thumbs while it drew power from the outlet at a trickle, gaining five miles' worth of energy for every hour of charging (a 220-volt outlet with 30–50 amps, such as those used for tumble dryers, would deliver about thirty miles every hour).

This fretful condition is what people are talking about when they use the words *range anxiety*. Unlike Tesla's dealer fights and its setbacks with fires, range anxiety has plagued electric cars for decades.

The term *range anxiety* appears to have first been used in 1997 in a

San Diego Business Journal article about GM's EV1, but it started becoming popular in 2010, about the same time GM trademarked it for the ostensible purpose of "promoting public awareness of electric vehicle capabilities." When GM applied for the trademark in July 2010, the company was months away from putting its hybrid Chevy Volt on the market. A charitable critic might agree with GM's claim that it trademarked the term to neuter its efficacy when used to cast doubt on the usefulness of electric cars. A less generous critic might argue that public concern about range anxiety could have helped sales of the Chevy Volt, which offered electric propulsion backed by a range-extending gasoline engine. An even less forgiving critic might go so far as to suggest that GM, which would face consternating profit problems were electric cars to become too popular too quickly, arguably had cause to perpetuate and promote the notion of range anxiety™ as much as it could. Whatever the case, it's now moot. GM abandoned the trademark in 2011, and with the Chevy Bolt (not to be confused with the similarly named Volt), it finally has a long-range electric car to promote.

The very fact of GM's trademark application showed how much a threat people considered range anxiety to be for electric vehicles. "It's something we call *range anxiety*, and it's real," a GM spokesman said before the Chevy Volt launch. "That's something we need to be very aware of when we market this car." GM positioned the hybrid Volt as a car first and electric second, he said, because "people do not want to be stranded on the way home from work." (These days, it advertises the Bolt on its website as an all-electric sedan that lets you "do it all.") Such statements did nothing to help the cause of electric cars, which in most cases have more than enough range to cover a driver's daily needs. The average American driver, according to US Department of Transportation data, drives just thirty-seven miles a day. A 2016 Massachusetts Institute of Technology study of American driving habits found that the Nissan Leaf, at seventy-four miles per charge, could replace 87 percent of cars on the road.

The effects of cold weather on the performance of electric cars' batteries exacerbate range anxiety. John Broder's failed winter Model S test drive for *The New York Times* in February 2013 perpetuated a notion that electric cars weren't suited to cold climates. In the article, he described turning the heater down and reducing speed in an effort to slow the drain on the battery, then driving with freezing feet and white knuckles. Similar fears about Tesla's cars suffering in cold weather were circulating elsewhere in the media at the time. *Consumer Reports* noted that its Model S suffered reduced range in cold weather, managing only 176 miles of driving despite the vehicle's digital dashboard saying it had 240 miles in store. On online forums, Tesla owners reported comparable effects, and a survey of a hundred electric-car owners by the research firm PlugInsights found that driving range in cars such as the Nissan Leaf and Ford Focus Electric diminished by between 25 and 50 percent in cold weather.

Frigid temperatures have two negative impacts on electric cars. On the one hand, chemical reactions in the battery happen more slowly in the cold, resulting in less current than at higher temperatures. On the other, the cold compels people to turn on their heaters, draining the battery further. Tesla has claimed that the Model S loses about 10 percent of its range in cold weather.

Results seem to vary for individual owners. Many Tesla owners in cold and snowy Norway say they are happy with their cars' winter performance, with one estimating that he lost between 10 and 20 percent of range for long trips in harsh conditions. On the other hand, a Tesla owner from Massachusetts analyzed the range of his Model S after driving about a hundred miles a day for a month in subzero temperatures and found that it diminished by roughly 40 percent.

Whatever the numbers said, Tesla's concerns about cold-weather performance were inseparable from concerns about range anxiety as it sought to convince car buyers everywhere that the Model S was a wise purchase. Tesla couldn't get by forever on friendly customers in temperate California. It needed to show that its cars could compete with gasoline vehicles in any conditions, even if it didn't enjoy the advantage of convenient fueling from the near-ubiquitous gas stations that had been built over the course of more than a century.

Skeptics have long said that lack of infrastructure meant that electric cars would struggle to find acceptance among the public. For example, a 2009 PricewaterhouseCoopers study suggested that inadequate infrastructure would restrict electric cars to short commutes. It was, as *Scientific American* concluded, "The Great Electric Car Quandary." Of course, skeptics said the same thing about Thomas Edison's electric light bulbs. Critics assumed that the market would continue to favor gas lamps, which were supported by a well-established infrastructure. Soon after Edison demonstrated his bulbs in a grand display at his Menlo Park lab in January 1880, a letter writer to *The New York Times*, who identified himself as F. G. Fairfield, PhD, of the New York College of Veterinary Surgeons, surmised "on practical and economical as well as on scientific and optical grounds, that the Edison system in its present state could not successfully compete with gas."

What happened with light bulbs, however, is that they became just one piece in a comprehensive new system that also encompassed generators, wiring, meters, and light switches—which were all part of Edison's plans. Josh Suskewicz, writing in the *Harvard Business Review*, characterized this system-level thinking as Edison's big breakthrough. "An electric lightbulb without ready access to electricity is a novelty; with it, it's a world changer," said Suskewicz.

In searching for an infrastructure solution for electric cars, Tesla struck upon the concept of a global Supercharger network. Fully refueling a Model S from a home charging installation takes hours because of limited power capacity, but a Supercharger does the job in under an hour by pumping up to 120 kilowatts of power directly into the car's battery. In about half an hour, it can deliver enough power to give the Model S 170 miles of range. Like the home charger, the Supercharger plugs into the Model S via an onboard port hidden by a trapdoor where a gas cap would otherwise be. Tesla's goal was to put Superchargers wherever people might want to drive their Teslas long distances.

Supercharging isn't quite as convenient as a five-minute gas station fill-up, but it is cheaper. For owners who bought their cars before January 2017, Tesla offered Model S and Model X road-trippers use of the chargers for free. For cars bought from 2017 on, Tesla offers four hundred kilowatt-hours of free charging per year (the equivalent of about a thousand miles), beyond which users must pay a small fee. Tesla has also made an effort to locate Superchargers near shops and restaurants so owners can be entertained while they wait for their cars to charge. They can check charging progress remotely with Tesla's smartphone app.

Tesla had secretly built Superchargers at six locations in California when Musk announced the technology at a party outside the company's Hawthorne design studio on September 24, 2012. Musk promised to cover most of the United States with Superchargers within two years, and the entire country, as well as the lower parts of Canada, within five. The crowd at the event didn't seem to know quite what to make of the announcement, perhaps because Musk delivered it in such a halting manner that—despite a smoke-and-lights show—somehow came across as understated. The fans in attendance offered only muted applause.

Musk added that Tesla would build solar canopies over the Superchargers so that the sites would generate more power than the cars used. (So far, Tesla has failed to fully live up to that promise—only a few Supercharging stations today have solar canopies.) Still, the crowd responded limply. Musk had seemed ready for a rapturous reception that was not forthcoming. He considered this a historic day, on par with SpaceX docking with the International Space Station earlier that year. But few seemed to grasp the magnitude of what he was promising. The next day, the press gave the announcement only tepidly interested write-ups.

Musk, however, did have one knockout sound bite from the night, and it promised a glorious escape from the oil-benighted status quo. He hit on the concept of freedom, noting that people thought the limited range of electric cars restricted them to short-distance travel. Not only would the Superchargers fix that, but they would also open up new electric highways. Coupled with solar power, Model S owners could liberate themselves from the worst effects of burning gasoline. And so Musk uttered a line crafted to excite the imagination. "You'll be able to travel for free, forever, on pure sunlight!"

By late January 2014, Tesla had completed the construction of a cross-country Supercharger corridor that would allow Model S drivers to get from Los Angeles to New York without having to spend a penny on energy. The electric highway took a northern route through Colorado, Wyoming, South Dakota, Minnesota, and Illinois, before approaching New York from Delaware. The path it cut was similar to a trip taken by Musk and his brother, Kimbal, in a beat-up 1970s BMW 320i in 1994.

Within days of the route's completion, Tesla staged a cross-country rally to show that the Model S could easily handle long-distance driving, even in the dead of winter. Two hot-pepper-red Model S's, driven by members of the Supercharging team, left Tesla's Los Angeles–based design studio just after midnight on Thursday, January 30. Tesla planned to finish the trip at New York's City Hall on the night of February 1, the day before Super Bowl XLVIII, which would take place at MetLife Stadium in East Rutherford, New Jersey, just across the state line. Along the way, the cars would drive through some of the snowiest and most frigid places in the country, in one of the coldest weeks of the year.

The trip took a little longer than expected. The rally encountered a wild snowstorm in the Rocky Mountains that temporarily closed the road over Vail Pass and then provided an icy entrance to Wyoming. Somewhere in South Dakota, one of the rally's diesel support vans broke down, forcing its occupants to catch a flight from Sioux Falls to rejoin the rest of the crew in Chicago. And in Ohio, the cars powered through torrential rains as the fatigued crew pressed on for the final stretch.

It was 7:30 A.M. on Sunday, February 2, when the Teslas rolled up to New York's City Hall on a bright, mild morning. The 3,427-mile journey had taken 76 hours and 5 minutes—just over three days. The cars had spent a total of 15 hours and 57 seconds charging along the way, which was good enough to establish a Guinness World Record for least nondriving time for an electric car traveling across the United States. Tesla calculated that each car saved about $435 on fuel costs by not having to pay for gas.

The cars had arrived 111 years after a thirty-one-year-old doctor named Horatio Nelson Jackson and a twenty-two-year-old bicycle mechanic named Sewall K. Crocker completed the first cross-country drive in a "horseless carriage." There were no interstates in those days. In fact, fewer than 150 miles of the country's roads were even paved. Jackson had decided to embark on the journey in the spring of 1903 after accepting a fifty-dollar bet from friends in San Francisco that he couldn't make it to New York in his cherry-red Winton touring car. He wanted to prove that the automobile, in the words of the narrator from the Ken Burns documentary *Horatio's Drive*, "was more than a rich man's toy, suitable only for short drives on city boulevards."

Range anxiety was a concern for Jackson and other owners of the horseless carriage, as it would continue to be for many years. At one point on their trip, with the Winton stranded, Crocker had to bike twenty-six miles to get gasoline from the nearest town—and then walk back after one of the bike's tires was punctured. The first drive-up gas station in the United States didn't arrive until ten years after the men made their journey, and five years after the introduction of Henry Ford's Model T.

On their transcontinental crossing, Jackson and Crocker had to drive through streams and over mountain roads that weren't designed for cars. They moved boulders by hand, endured thirty-six hours without food after getting lost in Wyoming's badlands, and got stuck in a swamp that buried the car up to its floorboards. The Winton survived a broken clutch, a clogged oil line, and a leaky gas tank, among a litany of other ailments. But at 4:30 A.M. on July 26, the two men, accompanied by a bull terrier they had collected along the way, drove up an empty Fifth Avenue in Manhattan in quiet triumph. The 4,500-mile road trip had taken sixty-three days and cost $8,000. But they got it done.

By comparison, Tesla's Cross Country Rally had it easy. The pair of gleaming red electric cars completed their run into New York with a crossing of the Brooklyn Bridge at sunrise. The mountains were far behind them, the ice had fallen from their fenders, and the skies had been wiped clean. A shaft of sunlight shone through the towers of Manhattan and illuminated the road ahead.

Fires could not end Tesla's hopes from there, and it seemed inevitable that its fights with auto dealers would do little to halt Musk's momentum. And now Tesla had struck an important blow against range anxiety, overcoming one of the last remaining obstacles to the proliferation of electric cars. Through sheer determination, Elon Musk, like Horatio Jackson, had set the scene for a new era in the history of transportation.

PART TWO
POWER SHIFT

7

GO AHEAD, START
A CAR COMPANY

"Why are we doing these things?"

M usk stepped onstage at Tesla's annual shareholders' meeting on May 31, 2016, and announced a departure from normal procedures. He was dressed in jeans and a casual blazer with cargo pockets. Soon he was joined by JB Straubel, who wore a white shirt and suit pants. This was fairly normal business garb for the cofounders, but the contents of their presentation differed from what one would usually hear at an event held to meet a public company's fiduciary responsibilities. They were about to talk the audience through the history of Tesla.

"I think it's important to explore the history and the motivations and the decisions along the way," Musk said, "so that people understand: What is Tesla all about? What does Tesla mean? Why are we doing these things?"

The CEO would do most of the speaking over the course of the presentation and keep the mood light, interspersing the narrative with self-deprecating jokes about the challenges Tesla has faced over the years. "We had no idea what we were doing," Musk recalled of the company's early days. "Like, how the hell do you build a car?" The crowd laughed with him. "No idea."

Musk started by describing his first meeting with Straubel, over lunch at a seafood restaurant near SpaceX. Straubel had come with Harold Rosen, an engineer with whom he had been working on an electric airplane. Rosen had brought Straubel to help pitch Musk on investing in the plane project, but Musk wasn't interested. Then Straubel mentioned that he had been discussing an electric-vehicle project with some friends from Stanford who had built a solar-powered car. Their idea was that they could strap together thousands of 18650 lithium-ion battery cells, which looked like AA batteries, to power a car for hundreds of miles. *That* was a plan Musk could get behind.

Musk had known since he was an undergrad that he would devote part of his career to electric cars. At the University of Pennsylvania, he had studied ultracapacitors for use as an energy storage mechanism for electric vehicles and further pursued the interest during two summer internships in the early 1990s at a Silicon Valley energy storage company called Pinnacle Research. Advanced energy storage was also going to be the subject of the PhD he had planned to do at Stanford before he ultimately decided to start Zip2 instead. Two Internet start-ups later, he was ready to turn his attention back to his electric-car passion. Indeed, two decades since he attempted to charm the young Christie Nicholson by inquiring if she, too, daydreamed about electric cars, Musk had finally found a partner who could indulge his predilection for sustainable transport.

After the lunch, Straubel sent Musk an e-mail seeking investment for the project in an attempt to raise $100,000. "It truly is a fascinating project and could have a phenomenal impact on the public perception of EV range and viability," Straubel wrote. "It is also a great way to keep a whole new generation of engineers interested and educated in renewable energy and efficient vehicles by building something that can really have an impact on the field. I feel quite strongly that the future of electric transportation will use high-energy density batteries instead of fuel cells and this is one step toward demonstrating that." Musk committed $10,000.

Soon after, Straubel introduced Musk to the head of AC Propulsion, Tom Gage, who was a close friend of the young engineer. AC Propulsion had developed an all-electric sports car called the tzero, which had a lithium-ion battery pack, boasted a range of 300 miles, and could accelerate from zero to sixty miles per hour in under four seconds. After test-driving the tzero, Musk tried over the course of several months to convince AC Propulsion to commercialize the car. But Gage and company weren't interested. Instead, they planned to produce an electric version of the Toyota Scion. Musk decided to move forward with an electric car company anyway. Gage offered to introduce him to a guy named Martin Eberhard, who had a similar idea.

In 1997, Eberhard and his friend Marc Tarpenning had started a company called NuvoMedia, which developed an electronic book reader called the Rocket eBook, a precursor to Amazon's Kindle. After a couple of years and sales of tens of thousands of units, they sold the company in January 2000 to Gemstar, *TV Guide*'s parent company, for $187 million. Like Musk with Zip2, they timed their exit to perfection, cashing in just before the dot-com bubble burst.

Eberhard and Tarpenning were neighbors in Silicon Valley's wealthy Woodside neighborhood, a twenty-minute drive from Palo Alto. After selling NuvoMedia, they cast around for other start-up ideas, seeking to embark on a venture that could make a meaningful contribution to the world. At one point, they considered developing an elaborate irrigation system for farms and homes based on a smart water-sensing network, but it was the tzero that ultimately caught their imagination, like it did for Musk.

Eberhard, who was growing increasingly worried about global warming, saw potential for commercialization and a chance to show that gasoline wasn't the only answer for motor vehicles. At the same time, he and Tarpenning had noticed that lithium-ion batteries had been improving at a rapid rate and were getting cheaper, thanks largely to their use in laptop computers. The auto industry, too, no longer seemed as impenetrable as it once was. Since the 1990s, automakers had been outsourcing many aspects of vehicle production, including the sourcing of components and in some cases even assembly. The men figured it would be possible for a start-up to design and build at least a prototype, with the hope of later raising more money to advance their ambitions. If things went well, a low-volume electric sports car with kick-ass acceleration might just give them a tiny foothold in the trillion-dollar auto industry.

By the time Musk was introduced to them, Eberhard and Tarpenning, along with a third friend and neighbor named Ian Wright, had incorporated a company they called Tesla Motors, an homage to Nikola Tesla. The company had the outline of a business plan but no prototype, no IP, and no funding. Soon after meeting in April 2004, Musk agreed to invest $6.35 million in Tesla Motors' Series A funding, out of a total of $6.5 million. Musk became chairman of the board and worked extensively on technology, product, and growing the public's awareness of the company, while Eberhard took on its day-to-day operations. Musk also convinced Straubel to join the company.

Over the next two years, Tesla recruited engineers from Lotus, Stanford, and around Silicon Valley to join their fledgling start-up. The first mule was ready in November 2004, and Straubel, who had become the chief technology officer, was given the honor of the first drive. It had taken the company just three months to go from the first schematics to a functional car. The result was a bare-bones vehicle with no body panels, a new battery pack, and the insides of a prototypical Tesla stuck on the chassis of a Lotus Elise. Straubel tore off down the road outside

Tesla's new office in San Carlos, six miles from Menlo Park, as his co-workers stood around in awe of their creation.

Drew Baglino, an early engineer at the company, also took a spin. At the 2016 shareholders' meeting, he recalled what it was like. "It was my first four-second zero-to-sixty experience, and I had never experienced anything like that," he said onstage after being invited up by Musk. "My prior car was, you know, an eighty-horsepower [Honda] Civic or something like that." The car, to everyone's surprise, held together. "It was an amazing four seconds, and that certainly hooked me on electric." Baglino would rise to become a vice president of technology at the company. "I haven't looked the other way ever since."

Musk drove the mule, too, and had enough confidence in its abilities to sink another $9 million into Tesla as part of a $13 million Series B funding round, which also included Valor Equity Partners. Over the next year and a half, Tesla developed an engineering prototype that was much closer to a production-ready vehicle and, in a stable economic climate that was also buoying the rise of emerging tech giants such as Facebook and YouTube, raised another $40 million. Musk contributed $12 million in the new round and, after a ten-mile-an-hour test drive (there was a bug in the software), convinced friends Larry Page and Sergey Brin, the founders of Google, to join in, too. Joining them in the round were some high-profile investment firms, including Draper Fisher Jurvetson, VantagePoint Capital Partners, and J.P. Morgan Securities.

Even when the Roadster was unveiled to the press in July 2006, three years to the month from the company's founding, few people had heard of Tesla Motors. Eberhard, who wore thin-framed spectacles and a tidy salt-and-pepper beard, fronted the event and took reporters on test rides, enthusing about the possibilities of an electric future.

"If you took the energy in a gallon of gas and used it to spin a turbine, you'd get enough electricity to drive an electric car 110 miles," he told the reporter who had the first ride that day. Eberhard described the electric cars the world had seen to date as "punishment cars" that were slow, unattractive, and cramped. The Roadster, meanwhile, looked like a classic sports car, something a millionaire motorhead would be happy to own. The only thing missing would be the purr of the engine—but Eberhard had a sound bite to excuse that, too. "Some people are going to miss the sound of a roaring engine," he said, "just like people used to miss the sound of horse hooves clippity-clopping down the street." Tesla would start shipping the Roadster in the summer of 2007, Eberhard said, and it had started work on a four-door sedan.

In the background, however, things were getting messy. A transmission that came from a supplier didn't work. A factory in Thailand proved unsuitable for manufacturing battery packs. Suppliers wouldn't pick up the phone, let alone deliver parts for a niche car of uncertain marketability. Paint wouldn't stick to the car's carbon-fiber panels. After being forced to change the transmission, the engineers also had to redesign the motor.

Following an audit by Valor Equity, one of the company's investors, Musk and Tesla's board found that the Roadster would be massively over budget and couldn't be made in time for the planned September 2007 launch. Eberhard ascribed blame to Musk's micromanagement and continual demands for changes to the car's design. Musk blamed Eberhard's mismanagement and a claimed lack of understanding of the financials. Meanwhile, as the summer of 2007 came and went, customers who had written checks to reserve their Roadsters started to ask, with increasing annoyance, what was taking so long. Musk and the board decided to act on a plan to relieve Eberhard of his CEO responsibilities and instead have him assume the title of President of Technology. In August, after the company had conducted a months-long search for a new CEO, Musk called Eberhard to tell him that Michael Marks, an early Tesla investor and the former CEO of electronics services company Flextronics, would be taking over as interim CEO. Eberhard, who had agreed to these plans, stepped down and assumed his new role. Three months later, though, Eberhard resigned in unhappy circumstances. He later said he didn't feel he had a choice. Tarpenning also left.

Musk and Eberhard traded insults in the ensuing press coverage. Eberhard told *Fortune* that he had no issues with Tesla as a company but he did "have problems with Elon and the way he treats people." Musk didn't respond warmly. "I was too busy trying to fix the fucking mess he left," he told the magazine, explaining why he hadn't yet told his side of the story. "I will say, I have never met someone who is as capable of creating such a disinformation campaign as Martin Eberhard." In November 2008, Eberhard told *Newsweek* that he thought Musk was a "terrible CEO." Musk responded by saying that "Martin is the worst individual I've ever had the displeasure of working with."

The dispute culminated in a legal tangle. In May 2009, Eberhard sued Musk and Tesla for slander. Musk fired back, first with a blog post that itemized Eberhard's alleged missteps as CEO, and then with two motions asking a judge to dismiss the suit. In September 2009, the parties announced that they had settled the suit. Each man issued a conciliatory statement about the other. Musk said, "Without Martin's indispensable efforts, Tesla Motors would not be here today." Eberhard said, "Elon's contributions to Tesla have been extraordinary." The confidential settlement also resulted in Eberhard's announcing that Tesla had five official cofounders: Eberhard, Musk, Straubel, Tarpenning, and Wright.

Marks, the new CEO, instituted cost-saving measures, bringing much of the production work to San Carlos, and pursuing a strategy to prepare Tesla for sale to a large automaker. But Musk's vision for Tesla was much grander than being an electric-car subsidiary at an incumbent automaker. In 2001 and 2002, Musk, as the largest shareholder of PayPal, had argued against selling the company to eBay, believing that it could become much more than a facilitator of online payments. He thought it could replace many aspects of traditional banks. Similarly, he was convinced that Tesla sat at the center of an opportunity that was bigger than the Internet. It is impossible to believe that he would be content for Tesla to meet a PayPal-like outcome.

Tesla replaced Marks in December 2007 with Ze'ev Drori, an experienced operations-focused executive. Drori's career had taken him from the Israeli military to IBM to Fairchild Semiconductor before he founded his own semiconductor company, Monolithic Memories, which went public in 1980 and merged with Advanced Micro Devices in 1987. Drori later took a controlling interest in (and became president of) Clifford Electronics, a car alarm manufacturer, before selling it to Allstate Insurance in 1999. He was an accomplished race car driver who had competed in the Grand Prix of Long Beach, but he had no auto manufacturing experience, which seemed significant given that his primary task was to get the Roadster out to market as soon as possible. He also had trouble establishing his authority. Musk was making all the major decisions. People inside the company came to see Drori as a mere implementer of the chairman's will.

Drori was at the head of the company when Tesla held a small event at its San Carlos headquarters to mark the delivery of the first production Roadster, in February 2008. Anyone outside the company wouldn't have guessed. Musk took ownership of the Batman-black car and delivered a speech promising that Tesla wouldn't stop until every car on the road was electric. Even amid these celebratory scenes, however, he hinted at undercurrents of financial turmoil.

"So far, this is a very expensive car," Musk said with a laugh to the makers of *Revenge of the Electric Car*, a documentary that had been following Tesla's ups and downs. "Call this the $50 million car," he joked. "That's about the amount of money that I've invested in Tesla."

Fisker Automotive went out of business in 2013. It made one model, the much-hyped hybrid Fisker Karma, and delivered two thousand cars to owners before experiencing a raft of problems. The software was faulty. Parts broke. The cars were in constant need of complicated repairs. The company missed multiple production deadlines. Henrik Fisker quit his own company, and 75 percent of the staff were laid off when, five years after it unveiled the Karma, the company started considering bankruptcy. Fisker Automotive lost money on every vehicle and in its short life burned through hundreds of millions of dollars, including $192 million of a Department of Energy loan. It was ultimately bought by Chinese auto parts producer Wanxiang Group, which in 2015 renamed the company Karma Automotive and planned a relaunch of the flagship car, given the new name Karma Revero.

Coda Automotive also filed for bankruptcy in 2013. It produced an economical electric car with eighty-eight miles of range and a top speed of eighty-five miles per hour. Aimed at budget-minded car buyers, the Coda was built with the body of the Hafei Saibao, a cheap sedan made in China, and a power train installed in California. The Los Angeles–based company was headed by former GM-China president Phil Murtaugh, but its cars were too expensive for their quality, selling at the price point of an Audi A4 but with the performance of an outdated compact car. Coda opened four dealerships in California but sold just a hundred cars in a year. After filing for bankruptcy, it switched its focus to energy storage. In 2016, the cash-strapped venture's assets were acquired by a small energy company called Exergonix.

The DeLorean Motor Company went into receivership in 1982. It had been started in 1975 by John DeLorean, who had been the youngest-ever GM executive. Some industry folks credited him with creating America's first muscle car, the Pontiac GTO. DeLorean raised tens of millions in venture capital, including investments from Bank of America, singer Sammy Davis Jr., and the *Tonight Show* host Johnny Carson. Supported by incentives and investment from the British government, the company established a manufacturing plant in Northern Ireland. After production delays and cost overruns, the factory opened in 1981 but was staffed by inexperienced workers. Many of the resulting cars had problems with panels, alternators, and their distinctive gull-wing doors. The DeLorean DMC-12 looked like a work of science fiction, with its showy doors, stainless steel body, and rectangular hood. Unfortunately, it was a flop. Despite being a two-door coupe, its zero-to-sixty-miles-per-hour time was a disappointing 10.5 seconds, and its $25,000 price tag (about $62,000 today) was too high for its performance value. The company was forced to cancel a planned stock sale in 1982, and then John DeLorean was ensnared in a Federal Bureau of Investigation sting and charged with attempting to smuggle $24 million worth of cocaine into the United States. Although he was later cleared, he was unable to raise further funding for his company. DeLorean filed for bankruptcy, killing 2,500 jobs and $100 million in investment. The DMC-12, however, lived on as a time machine in the *Back to the Future* movies.

Fisker, Coda, and DeLorean were not outliers. The historical trend for new car companies is one of quick decline soon after birth—and, ultimately, failure. Detroit Electric, Think Global, Aptera Motors, Vector, Tucker, Kaiser-Frazer, Duesenberg—the list is long and stretches back to the early 1900s. In fact, all but two significant American auto companies started in the last hundred years have succumbed to the auto industry's unceasing challenges. One was founded in 1925 by a guy named Walter Chrysler. The other is Tesla Motors.

On the list of nifty business ideas to try, starting a car company in the United States should sit right at the bottom. The engineering part—coming up with a new car design specific to your hopes and desires—might be cool, but that's where the fun stops. From there on in, it is all grind, psych-outs, and financial massacre.

The costs start adding up early. After you've designed a vehicle that you think people might like, you have to figure out how to build it affordably. To do that, you have to pay engineers and designers, as well as lawyers to make sure you're on the right side of the law and safety regulations.

Once you've cleared that hurdle, you'll want to start producing cars by the thousands, so you need to buy a factory, which will likely cost about a billion dollars. Then, you'll spend tens of millions of dollars on equipment to go inside it.

Even if you're planning to make only twenty thousand cars a year, you need to convince suppliers that they should work with you. This step is trickier than it might initially seem, because your company will have no history in the industry and will be operating with volumes that are unattractive to most suppliers. Good luck getting the good ones on the phone.

If you're a special case, though, and you make it to the next stage, you should be ready to start manufacturing cars to go to market. At this point, you need to come up with a process that allows you to quickly and flawlessly produce vehicles to keep up with demand. That done, you can expect to shell out more for assembly workers, and then the lawyers come back in to make sure you comply with the Highway Safety Act and other federal standards. Meanwhile, testing for durability, performance, efficiency, and aerodynamics will cost you millions more.

Next, think about your sales strategy. Are you going to work with dealers? Or will you build your own retail network and spend millions more to contest state laws that protect dealers as the middlemen for new car sales? And what are you going to do about service? You need a slush fund to cover warranties, recalls, and lawsuits, and you have to certify, train, and employ mechanics and technicians to deal with repairs.

Let's say your car start-up has defied the odds and is finally booming. Then it will be time to hire thousands more engineers, designers,
and factory workers. You might even have earned the luxury of thinking about advertising or other ways to build your brand. After all, you
need to differentiate your cars from the entrenched incumbents. So
take a deep breath and go back to your investors to ask for more
money—your costs to date would have cleared a couple billion dollars.

Now imagine doing all that, except instead of starting a normal
car company, you decide to bet on electric propulsion. That's a lot
harder. It means you have to create a lot of new technology and rein-
vent some of the design and manufacturing processes. You have to
make sure that owners of your cars won't easily run out of charge, and
that they can always find a place to power up no matter where they
are in the world. You have to teach people that electric cars can actually be pretty good, and that they should consider choosing something other than the automotive technology they've known their
entire lives. If you're even remotely successful in these endeavors, you
can expect skepticism, hesitation, and in some cases active disparage-
ment as you attempt to reach more customers. Get ready to do a lot
of explaining.

And most of all—this is really important—you should pray that,
just as you're getting started, the global economy doesn't suddenly go
into free fall.

-----~~~~~~~-----

Scene I: Late 2008, a group of managers meet in a Tesla conference
room. Elon Musk is thirty-six years old. His hair is shorn short. Beside him sits Deepak Ahuja, who has recently joined the company
from Ford to be Tesla's chief financial officer.

Musk: "We need to get the company to cash-flow positive in six to
nine months, or we're screwed."

He looks shell-shocked.

"It's really pedal to the metal here. Each month that passes literally costs us tens of millions of dollars. We need to appreciate that." He stares into space.

In the last two years, Tesla has burned through $100 million. It has made just over a hundred Roadsters. One of Ahuja's first tasks is to cut 30 percent from the Roadster's power train costs.

Musk is told that faults in some of the cars can be attributed to substandard parts. The engineers didn't realize until it was too late. Now he is pissed.

Musk: "I want names named. If someone's always on the hot seat and is always the root cause for problems, they will not be part of this organization long term."

Scene II: Not long after the meeting, Musk visits a Tesla vehicle delivery center in Menlo Park. Awaiting him is a workshop full of defective Roadsters.

Musk: "Holy mackerel!" His hands go to his head. "Jesus! We have, like, an army of cars here. Like, Jeeeeesus!"

Musk tells the team to overhire to fix the problem.

"I'm available twenty-four/seven to help solve issues. Call me three A.M. on a Sunday morning, I don't care."

Background: September 2008. In the face of a liquidity crisis set off by a collapse of subprime mortgage repayments, Merrill Lynch sells to Bank of America, Lehman Brothers files for bankruptcy, and the US Treasury takes over mortgage finance companies Fannie Mae and Freddie Mac. The stock market tanks, credit lines freeze up. The financial world is in crisis. It becomes near impossible to raise money in Silicon Valley.

Scene III: Dan Neil, auto columnist for *The Wall Street Journal*, is interviewed in the documentary *Revenge of the Electric Car*.

Neil: "Elon's going to lose his shirt!"

Background: As the financial crisis hits, the gossip blog *Valleywag* reports that Tesla is down to its last $9 million. The company delays its plan to raise $100 million. It pushes back the start date for building the Model S, even as Franz von Holzhausen, hired from Mazda, has started work on its design. Musk takes over from Drori as CEO and lays off a quarter of the company's 360-odd employees.

Meanwhile, Musk is going through a divorce with Justine, who's documenting the experience on her personal blog, and SpaceX is also in financial dire straits after its first three launch attempts fail.

Scene IV: Sometime in 2014. Musk, in an interview with the biographer Ashlee Vance, reflects on that period in 2008.

Musk: "I thought things were probably fucking doomed."

What happened at Tesla as it pulled itself back from the edge of a cliff over the next five years is one of the reasons Musk is revered in the tech business. The legend stems in no small part from what happened on Christmas Eve 2008, when the electric revolution was days away from a premature end. Two days earlier, Musk had learned that SpaceX had won a $1.6 billion contract from NASA to supply the International Space Station. ("I couldn't even maintain my composure," he would later say about the fateful phone call from NASA. "I was like, 'I love you guys!'") Musk had pleaded for more funds for Tesla from existing investors and put in $20 million of his own, which he scraped together from various sources, including his proceeds from the late-2007 sale of his cousins' data center start-up, Everdream, which was acquired by Dell. His friend Bill Lee, an entrepreneur and investor, wrote a check for $2 million, and Sergey Brin put in half a million. Several Tesla employees contributed in increments of $25,000 and $50,000. At 6:00 P.M. on December 24, Musk finalized a financing round of $40 million. It would be enough to keep the lights on just a little while longer.

By March the next year, von Holzhausen and his design team had rallied to produce a Model S show car. This was achieved by way of mayhem. Even as the car was waiting to be shown to the invited guests at the SpaceX factory, the team was fiddling with its parts and fixing the seats. Between test drives, the team pumped ice water through parts of the power train to keep it from overheating. Some of the panels were stuck to the frame with magnets.

The gambit somehow worked. Reporters praised the vehicle, with *Wired* calling it "stunning" and *The New York Times* comparing it to a Maserati. Musk, though, was exhausted. "The last few days, last few months, [have] just been a very, very high workload," he told a film crew that had been following his progress. "Not enough time to sleep."

Why, indeed, was he doing these things?

———〰〰〰〰———

At the shareholders' meeting in 2016, Musk and Straubel identified Tesla's savior: a German automaker that would one day become a rival.

In late 2008, Musk met with Dr. Thomas Weber, a Daimler board member, who told him the company wanted to make an electric version of its Smart microcar, but it didn't have a good source for the battery or power train. A team of senior Daimler engineers would be visiting Silicon Valley in January 2009, Weber told Musk. "I was like, 'Okay, wow,'" Musk recalled. "As soon as I left that meeting, I called JB and said, 'We have three months to make a working electric Smart car.'" Straubel called Musk's idea a "non sequitur," because Tesla was hard at work trying to make the Roadster, which was causing more than enough problems on its own. At the time, you couldn't even buy a Smart car in the United States. But Musk saw an opportunity to attract a powerful partner.

Tesla sent someone to Mexico to get a Smart car and drive it back to California. When it arrived at the headquarters, a small SWAT team of engineers tore out its propulsion system and started designing a new battery pack for the one-off project. Musk, as usual, issued difficult directives: The car needed to look unmodified, and the power train couldn't encroach on the passenger compartment. "That team didn't sleep a whole lot in those couple of months," Straubel said. The engineers figured out how to adapt the Roadster's motor and power electronics system to the new vehicle and soon realized that it was going to be the fastest Smart car ever made. It would have all the torque of a Roadster in a compact machine fewer than nine feet in length. "It was so fast," Musk said, "you could do wheelies in the parking lot."

When the Daimler engineers showed up at Tesla, they were not excited to meet some random American car start-up. Tesla's executives started with a PowerPoint presentation that wasn't well received, before Musk interjected to suggest going straight to a test drive. Daimler's engineers didn't know what he was talking about—as far as they knew, there was no electric Smart car. "We made one," Musk told them. "It's just outside. You want to drive it?" Soon, the Daimler emissaries were on the road in an insane-performance Smart car. "They went from being a bit grumpy," Musk said, "to 'Holy cow, this is awesome!'"

That test drive resulted in a development contract that Musk now credits with saving Tesla. The German giant enlisted the start-up to make the Smart power trains, and in May 2009, it announced it was buying 10 percent of the company for $50 million.

The Daimler deal, as well as the Model S prototype unveiling, helped secure a $465 million loan from the US Department of Energy, part of a George W. Bush administration–led program to encourage the development of alternative-energy vehicles. The deal required Tesla to build electric cars and power train components in the United States and pay off the loan within twelve years. While the loan was confirmed in 2009 during President Obama's first year in office, the first payments weren't due until early 2010. Now Tesla would have enough money to get started on the Model S in earnest. Roadster sales were also bringing in much-needed revenue.

Tesla got another $50 million and a hugely important partnership when, in May 2010, Toyota bought a 2.5 percent stake in the company and contracted it to build power trains for the electric version of the RAV4. But Toyota brought something even more important to Musk: a factory in Fremont, California.

Securing a factory had been a problem for Tesla. It had already had two false starts with proposed factory locations in Albuquerque, New Mexico, and San Jose, California. But here was an elegant solution to its manufacturing question, lying just across the bay from its Palo Alto headquarters.

Toyota originally wanted more than $100 million for the five-million-square-foot facility, which it had run with GM under the auspices of New United Motor Manufacturing, Inc. (NUMMI), but Tesla offered only $42 million. After a breakfast at Musk's house in Bel Air, Toyota CEO Akio Toyoda told his company to accept the offer.

Things were looking up, and Tesla seemed poised for acceleration. But even with a fresh hundred million in the bank and a big loan from the government, life wouldn't be easy. To meet the expectations it had set for itself, Tesla would have to build twenty times as many cars as it had so far sold—all in the space of twenty-four months. By mid-2009, it had burned through $300 million since its founding and would soon need a lot more money to equip the factory, hire thousands of employees, and open more stores. That meant Musk would have to do something he had never done before: take a company public.

With Zip2, PayPal, SpaceX, and, until that point, Tesla, Musk had followed a fairly conventional fund-raising strategy. He and his cofounders financed his companies with money invested by venture capitalists, investment banks, and wealthy individuals, all of whom hoped that the businesses would one day turn major profits or have a lucrative "liquidity event" that would result in huge payoffs. One such liquidity event is an initial public offering (IPO), wherein a company lists its stock on an exchange so that members of the public can buy shares, too. At an IPO, existing shareholders have an opportunity to sell their shares at whatever the public market is willing to pay for them—in many cases a substantially larger sum than early investors would have paid when the company was privately valued. A company that goes public has to be more transparent about its financial practices and positions, but it can also benefit from a vast new cash base as investor dollars pour in. In filing to go public in June 2010, Tesla indicated that it hoped to raise up to $178 million in its IPO.

On June 29, 2010, the day of Tesla's initial public offering, Jim Cramer, host of the CNBC show *Mad Money*, advised viewers to steer clear of the stock. "You don't want to own this stock! You don't want to lease it! You shouldn't even rent the darn thing!" The company, he pointed out, had sold only 1,063 cars. Indeed, its listing price of $17 a share seemed ambitious, especially given that the NASDAQ was down 2 percent that morning.

Investors ignored Cramer's advice and rewarded the first American car company to go public since Ford with a booming first day. By the time the market closed, Tesla's stock had surged to $23.89, giving the company $226 million. On the NASDAQ podium at the closing bell, Musk, with his son Griffin in his right arm, lifted his left arm in triumph and grinned from ear to ear.

Tesla continued to bleed money in producing the Model S and took longer than expected to get the car to market. A year after the IPO, it sold another $158.5 million in stock, and in 2013 it found itself so low on funds that it reportedly considered selling to Google. In May that year, it sold a combination of stock and debt in a billion-dollar financing arrangement that allowed it to pay off its Department of Energy loan. The company marked the repayment, which came nine years ahead of deadline, with a press release in which it pointed out that none of the car companies that had received bailouts in the financial crisis had fully paid off their debts.

That May was a good month. The day after Tesla reported its first profitable quarter, *Consumer Reports* called the Model S the best car it had ever tested, awarding it a score of 99 out of 100. Four days later, new sales figures revealed that the Model S was the highest-selling premium sedan on the market, outstripping even Daimler's Mercedes S-Class. Daimler may have been starting to doubt the wisdom of bailing Tesla out four years earlier.

As Tesla checked off more achievements in 2013, the company's reputation took on a luster as rich as its car's paint job. In August, Model S deliveries started in Europe, and NHTSA gave the car its best possible safety rating. In November, just as struggling rival Fisker filed for bankruptcy, *Fortune* named Musk its Businessperson of the Year. In early 2014, *Consumer Reports* named the Model S its top pick for the year.

In February 2014, Tesla revealed plans for a battery factory that would make its Fremont plant look like a doll's house. This time, the company wasn't going to rely on luck or an advantageous friendship with Toyota. The Gigafactory, at which Tesla would partner with Panasonic to produce more lithium-ion batteries than all companies in the rest of the world combined, would be carefully planned from the start. Tesla would conduct a methodical nationwide search for the best site and seek financial incentives from the eventual host state. Then, according to initial plans, it would build the factory in phases. Now that Tesla had established itself as a viable business, Musk appeared to be building confidence that his wild ambitions were practical and achievable.

The milestones kept coming. In April, *Top Gear* magazine called the Model S the most important car it had tested, and Tesla started deliveries in China. The first right-hand-drive Model S was delivered to the UK in June.

Then, in a surprise move, Musk defied convention by declaring that Tesla would make its patents freely available to all. He promised that the company wouldn't initiate lawsuits against people who used Tesla's patents without paying—even competitors (provided they used them in good faith). Why would he do this thing? Most companies view patents as a way to protect their inventions and at least ensure that the ensuing license fees provide a reliable source of revenue, but Musk saw them as legal instruments that large corporations use to stifle competition. He wanted to make it easier for other automakers to build electric cars so that the world could benefit from reduced carbon emissions. Plus, the resulting attention could help with recruiting talented engineers, which he saw as key to Tesla's business.

Soon after the patents announcement, a Morgan Stanley analyst designated Tesla "America's most important car company." In September, after the Model S received yet another plaudit from *Consumer Reports*, Tesla's stock price hit $291, an all-time high. By the end of the year, there would be more than 70,000 Teslas on the road.

Tesla Motors had transformed from stuttering start-up to fledgling fairy tale. Musk had gone from bordering on bankrupt to tech-business hero. He was inspiring fans and followers, attracting large crowds at conferences, appearing on magazine covers galore, and, in January 2015, starring as a guest voice actor on an episode of *The Simpsons* that was built around his character. In May 2015, the nerd blog *Wait But Why* called him "the world's raddest man."

As Musk's fame spread, so too did the seeds of the electric revolution. The electric car, once dismissed as a fool's fantasy, was beginning to look like an enticing business opportunity. Starting a car company no longer seemed such a crazy idea.

Through an airplane window at ten thousand feet, I saw Las Vegas as a scene of environmental violence. The city sat on the flat bottom of a basin in the Mojave Desert, a beaten-down beige interrupted by rip-ples of rooftops and splashes of stark greens where golf courses had been implanted by surgical procedure. Man-made ponds huddled in tight clusters. A thirsty river struggled to cut a course through town.

The dominant view of the urban skyline was, of course, the strip of casino skyscrapers that juts upward along an artificial fault line: the Venetian, Mandalay Bay, Bellagio—monuments to markets of chance.

The city's railroads, once prominent, are gone now—unless you count the monorail that shuttles tourists to the back entrances of the Strip's casinos. Its dominant transportation infrastructure is a lattice-work of roads. To get around without relying on cars or buses requires a feat of imagination. The airport's rental car center is itself the size of a small airport. Gasoline is the city's blood.

On a hot, sunny day in April 2016, Nevada governor Brian Sando-val and a handful of executives from a new car company gathered in an air-conditioned pavilion erected on a flatland of rocks and brush thirty miles to the north of the city. The tent sat on a nine-hundred-acre land parcel in the Apex Industrial Park, above which fighter jets from the nearby Nellis Air Force Base streaked in pairs across a vivid blue sky. The men had gathered for the ceremonial groundbreaking for the electric car start-up's vehicle assembly plant, intended to be built over the next two years at a cost of $1 billion. About twenty re-porters from China were present for the ceremony and asked ques-tions of Governor Sandoval in a Chinese-language press conference in a side room.

The construction schedule and scale of ambition were enough to create the impression that this was an Elon Musk project. But the start-up hoping to build the manufacturing facility in the Nevada desert was a Tesla follower, not an offshoot. It was called Faraday Future. The new company was based in Los Angeles and funded by an elusive Chinese billionaire. Its executives included a few former

Tesla employees, one of whom was Dag Reckhorn, a former Tesla manufacturing director who filled a similar role for Faraday Future.

Reckhorn, a floppy-haired German with an LA tan and prescription glasses that darkened in the bright daylight, stood onstage inside the tent and told an audience of a couple hundred guests and media (including me) that Faraday's cars would be built with "extreme technology" and the factory would be specced to match. It would be environmentally friendly, futuristic, and world-class.

"Our aim is to complete a project that would traditionally take four years"—he leaned into the microphone and repeated "four years"—"and we want to do it in half the time." He waited a couple of beats for applause that didn't come and then leaned in even closer to the mic. "And we still want to do a great job." The audience hesitated for a moment before breaking into stilted applause.

Faraday Future had no CEO, nor a drivable prototype. It had been incorporated in early 2014 but didn't come out of stealth mode until July 2015, when it announced itself via press release. The public had been exposed to little more than its website, a handful of news clippings, and some vague marketing hype about a new approach to mobility.

Reckhorn played a twenty-second video that showed a 3-D rendering of the factory. As a gladiatorial soundtrack pumped up the adrenaline quotient, fast-moving frames showed gorgeous water features, languid palm trees, and solar panels atop a structure rich in glass surfaces and reflective black floors. Any self-respecting robocar would be happy here.

Under Governor Sandoval, Nevada had fashioned itself as America's most business-friendly state. Sandoval was also establishing a reputation as a booster for high-tech manufacturing. As well as Faraday's billion-dollar factory, the governor had landed Tesla's Gigafactory in northern Nevada. Over a hill from the Faraday site, Hyperloop One was building a track to test its realization of Musk's so-called "fifth mode of transport." Bigelow Aerospace, maker of the Bigelow

Expandable Activity Module—an inflatable space habitat better known by the acronym BEAM—had set up headquarters just down the road in North Las Vegas. Three days prior to Faraday's factory groundbreaking, SpaceX's Dragon spacecraft had delivered a BEAM to the International Space Station. The Falcon rocket that sent the BEAM into orbit subsequently landed itself back on an autonomous drone ship in the Atlantic Ocean.

Following Reckhorn, Sandoval stepped up to honor his new tenants. "A little over a year ago, none of us could have conceived or imagined this," said the square-jawed former judge, taking in the surroundings. "This doesn't happen in other states, I promise you that."

Just over the horizon was a reminder of why it was so important for Sandoval to attract industry and jobs to the area. The city of Las Vegas had endured a serious hangover after going on a subprime mortgage bender in the mid-2000s. Stephen M. Miller, a professor of economics at the University of Nevada at Las Vegas, had called it "ground zero for the Great Recession." When the mortgage crisis struck in 2008, Las Vegas was the worst-affected major city in the United States. By 2010, unemployment rose to a record-high 14.8 percent. In early 2016, Nevada still ranked third in the country for joblessness.

After half an hour of speeches from government officials and Faraday executives, the ceremonies moved outside so the delegation could be photographed with hard hats on and shovels in the ground. As Sandoval and Reckhorn grinned with shiny shovels in hand, the cameras trained their lenses on a placard behind them that bore an artist's rendering of the factory and the words THE FUTURE BEGINS HERE. The guests sipped from glasses of champagne as dust swirled around their feet.

I walked back across the hard ground to my rental car and thought of Tesla's struggles over the years. The challenges that lay ahead of Faraday were immense. Did these people know what they were getting themselves into?

At Tesla's 2016 shareholders' meeting, Musk joked about the some-times too-grandiose promises the company had made in its early days. "We say the things we believe even when sometimes those things we believe are delusional," he said, eliciting laughter from the crowd. By that time, Tesla had learned many hard lessons and lived to tell the story. Its shareholders, enjoying the spoils of a venture that had seen its valuation rise from zero to $30 billion in twelve years, could afford to laugh. But the joke concealed the pain Tesla had endured and that surely awaited its followers.

Tesla was six years old and had produced hundreds of Roadsters when it scored a move-in-ready factory for the bargain price of $42 million. But here was Faraday Future, an unknown company not yet two years old, with no functioning car to its name, spending a billion dollars to build a factory from scratch with a two-year deadline. Even by Las Vegas standards, it seemed an insane gamble.

8

CALIFORNIA DREAMING

"Oh, it's a Chinese Steve Jobs here!"

What the hell *is* that?"
Matt Burns seemed incredulous. The *TechCrunch* reporter, seated onstage alongside Nick Sampson at the Sands Expo convention center in Las Vegas, was looking at a screen on the wall behind him. It showed a picture of the FFZERO1, a supercar that Faraday Future had unveiled three days earlier at the Consumer Electronics Show (CES). The concept car—or, as its designer, Richard Kim, had described it, "a car of concepts"—looked like a horizontal Scud missile on wheels, with an abbreviated nose, a fighter-plane cockpit, and a plexiglass fin along its spine.

Burns laughed at his own comment, as did some people in the crowd who had gathered around *TechCrunch*'s sponsored CES stage. Sampson laughed, too. The senior vice president of product research and development was Faraday's primary spokesman, an engineer whose CV included Jaguar, Lotus, and Tesla, and who had cofounded the latest electric car start-up to capture the tech world's attention. Sampson didn't look entirely comfortable, leaning back in a faux leather armchair and holding a microphone near his mouth as if it were some sort of shield. In a gray blazer over a blue shirt, he was dressed formally by the standards of CES, one of the world's largest gatherings of consumer electronics retailers, enthusiasts, and reporters. Burns, unshaven and sporting an untidier version of the blazer-and-shirt look, made no move to help Sampson settle. He left a silence for the spokesman to fill. "That's, um, a representation of our company," Sampson offered in an affable southern English accent. "It's good-looking, it's fast, it's dynamic, and it's, uh, revolutionary."

"It looks like a Hot Wheels car," Burns retorted. "Is it supposed to look like a Hot Wheels car?"

"It's supposed to attract attention," Sampson returned, "and get people talking about us."

Mission dubiously accomplished.

It was January 8, 2016, six months since Faraday Future had announced plans to build a long-range electric car by 2017. The company had been named in homage to British scientist Michael Faraday, who in 1831 discovered that an electrical conductor could interact with a magnetic field to produce an electromotive force. Little was known about the company except that it was based in Gardena, California, on the outskirts of Los Angeles, and had hired a slew of former Tesla employees, as well as Kim (a former BMW i8 concept-car designer) and Silva Hiti, former head of power train on the Chevy Volt. When *Motor Trend* magazine announced Faraday's existence, a spokesperson had said, "We're not Tesla. But we're not Fisker, either," referring to its short-lived predecessor. "We're not fucking around."

A series of reports since the initial media push had served only to increase the sense of mystery around Faraday. It was soon revealed that the company was funded by Jia Yueting, the founder and CEO of Chinese electronics and Internet media company Leshi, which in turn is the holding company for LeEco, a group of companies that produce and distribute smart devices, such as TVs, cell phones, and bicycles. (It can be confusing for casual observers, but LeEco was previously most commonly known as LeTV, Leshi's streaming-video unit, which, even more confusingly, rebranded as Le.com.) Faraday was planning to produce the cars in the billion-dollar factory that it would build in North Las Vegas. Sampson had told reporters that the company's business model wouldn't rely only on vehicle sales but would be based on subscriptions, apps, car sharing, and "other opportunities." The company would first launch a single model and then follow that with a range of others.

Public intrigue about the media-branded "Tesla competitor" had been sufficiently stoked by the end of 2015 to generate substantial interest in what it would reveal at CES. Faraday duly hyped itself with a series of posts on social media that featured teaser images and videos, promising "a glimpse into the future." When, amid pulsating dance music and a light show fit for a pro wrestling match, it lifted the cover on a flashy supercar that was clearly *not* the kind of vehicle the company had suggested it would bring to market, many industry observers were disappointed. *Bloomberg* derided it as a "1,000-horsepower Batmobile," declaring: "This is not the concept car you've been waiting for."

Burns, a gadget reviewer based in Greater Detroit who also covered *TechCrunch*'s auto beat, continued the skeptical tone. "So will this car ever be built?"

"Well, it is built, because there is a picture of it here," Sampson replied. "You mean a working one?"

"Yeah, will it ever work?" Burns's tone was so condescending that it verged on hostile. Concept cars—even ones that don't move—are a staple of the auto industry, and it would be reasonable to give a start-up that had revealed itself to the public only seven months earlier a little slack. Then again, Faraday had been aggressive in attracting media attention before it had anything real to show to the public.

Sampson continued. "Well, one of the hallmarks of Faraday Future is that we're a little bit mysterious and secretive about what we're doing." He shrugged his shoulders and, in an attempt at breeziness that fell just on the wrong side of awkwardness, added, "So just wait and see." Burns was undeterred.

"So tell me—why unveil your company with something like this rather than something practical or logical?" Sampson's response again wasn't entirely convincing.

"As I said earlier—we're a blend between a consumer electronics company and a car company." This blend was easier claimed than achieved. To date, Tesla, Sampson's former employer, was the only company that had effectively integrated the disparate cultures of the tech and automotive industries at scale.

"What does that *mean?*" Burns asked. "Because Ford will tell you that they're an automaker, a technology automaker. So you say you're a technology company rather than a car company." He gestured at the screen behind him. "That's a goddamn car. So why are you a tech company?"

"Because what we're, ah—the way we're looking at the, uh, what the user wants, what the user needs," Sampson responded, "as we look into the future. What's this show all about? It's about consumer electronics, it's about our digitally connected lifestyle." Burns looked at his notepad and adjusted it on his lap. A man in the crowd got up to leave. "And at the moment, the car industry isn't meeting those needs. Yeah, some companies claim to have got some connectivity, but it's just a pretty poor amount of connectivity, and it's not seamless with the rest of your lifestyle." Sampson noted that most people preferred the maps on their smartphones to onboard satellite navigation systems, and that few people could read their e-mails while sitting in their cars without getting their phones out. Music-playing options were inflexible.

"But that's only a tiny part of an automobile," Burns insisted. "That is the user experience and an infotainment screen, where in a vehicle you have drive systems, suspension systems, braking, safety, chassis design—and you know a lot about chassis." Burns seemed to suddenly remember that he was talking to the former chief engineer of Lotus. "So you're talking about a tiny portion of it. So are you focusing on that? Are you trying to sell that part of it? Does nothing else matter?"

Sampson responded, "That's why we're making the point of being a tech company and that we look at it from the consumer electronics side."

"All right," Burns said quickly, shutting down the response to his lengthy question. But at least he didn't get up and leave.

Six weeks later, *The Guardian* reported that Faraday had initially intended to reveal a working prototype of its planned production car at CES, but it had fallen behind schedule. It wanted instead to show a drivable version of the FFZERO1, but it fell behind on that, too. Ultimately, Faraday had little choice but to reveal the nonmoving "car of concepts," which reportedly cost $2 million to produce.

The reason Faraday could afford to build such an expensive non-car could be traced back to a rural town in China's northern Shanxi province. For Faraday to get started, Jia Yueting had to get rich.

Like many of China's self-made businessmen, Jia had humble beginnings. The third child of a teacher and a housewife, he graduated from the University of Shanxi with a master's degree in business administration and, as a twenty-two-year-old, worked for a year as a tech support officer in a county tax office. He soon left to start his own tech consulting firm and, according to Chinese newspaper articles, was later involved in various businesses: coal, restaurants, private schools, and selling batteries for cell-phone-tower antennas.

Jia's first major entrepreneurial success came in 2002, when he established the Sinotel group, a wireless communications provider that started off distributing cell phones in China and later expanded to designing, installing, and maintaining network infrastructure for China Mobile, China Telecom, and China Unicom, among other telco customers. Jia started LeTV in 2004 but remained as Sinotel Technologies' executive chairman as it listed on the Singapore Exchange in 2007, and then through its declining fortunes in the 2010s. The company's 2014 annual report showed a full-page photo of a smiling Jia dressed in a dark suit with his hands stuffed in his pockets. The report carried bad news: "We continue to experience a slowdown in our core business, compounded by the slower collection from the telcos." In February 2016, Sinotel announced it was delisting.

For the most part, business was looking better at LeTV, which went public on the Shenzhen Stock Exchange in 2010, listed as Leshi, the name of a holding company Jia also set up. LeTV's streaming service had acquired the rights to popular TV shows and movies and established itself as one of the most visited online video sites in China. Leshi later started a film production division, Le Vision Pictures, which partnered with acclaimed directors and invested in and distributed *The Expendables 2*, among other Hollywood movies. In 2014 and 2015, Leshi started selling smart TVs and smartphones. By 2016, it had a market valuation of about $14 billion and rebranded itself into two groups: Le.com for media services and LeEco for electronics products (its future cars would fall under this group). The name LeEco reflected its new positioning as an "ecosystem" of connected hardware products. Jia's personal net worth at the time of CES in 2016 was about $4 billion.

Back at CES, Sampson suggested to Burns that producing a car alone was just an entry point. "The car side, we've just got to be merely better than anybody else," he said with blithe confidence. "That's the easy part for a car guy like me. But what do the consumers want?" He considered a car's acceleration and top speed when making a purchase decision, but as a veteran of the auto industry, he was an atypical customer.

"You were talking to me like a car guy," Sampson continued, "but we're at the Consumer Electronics Show. Don't you want to know what the download speed is for this car? Whether it's 4G or 5G connected?"

Burns was unequivocal. "*No.*"

"Ah, I think some of the generations to come are going to be asking—"

Burns interrupted and looked to the crowd: "Does anybody want to know how fast it goes, or if it has 4G? How fast? Raise your hand." He paused to survey the crowd. Dozens of hands went up.

"4G—raise your hand." A few hands went up.

It was an odd way to frame the question. The high-speed 4G wireless connectivity standard has been around for years—even in Sinotel's heyday—and is now considered mundane. By contrast, 5G, which some experts estimate will be widely available in 2020, will be more than just a connectivity upgrade—it stands to be a phase-shifting technology for automotive by allowing fail-safe connectivity in its areas of coverage, and download speeds so fast and reliable that cars can communicate with one another, and with infrastructure, while on the move. The transportation system could become an Internet of its own, with vehicles turned into roving hotspots.

Sampson looked at the hands up in the crowd. "Quite a few," he said, sounding pleased.

"About *four*," Burns said dismissively.

He also asked Sampson if any parts of the FFZERO1 would be in the production car. Sampson replied that some of the styling features would, and that they were important to helping establish a recognizable brand. One of the key lines, he said, was a crease that ran around the car's midriff. "We call that the, uh, uh . . ." Sampson was struggling to recall the word. He laughed self-consciously. One of his colleagues in the front row prompted him. "The UFO line!"

The name was symbolic. "A UFO is obviously something out of this world," Sampson ventured. "It's something nobody recognizes or understands, and that's what we are."

On November 26, 2014, Jia announced on his Twitter-like Sina Weibo microblogging account that he had returned from surgery in Hong Kong (which the local media said was to remove a rare tumor called a thymoma) and had been in a Beijing hospital for further treatment. Jia's post provided an explanation for a months-long absence from the public eye in China. "No matter the hardships I've been going through, I'll never stop pursuing my dream of promoting humanity's progress," he wrote. He posted a photo of the view from his window, which showed a blurry sun lost in an inky sky. He vowed to promote a "new industrial revolution" to "improve the environment so that every Chinese person can breathe clean air."

In the same week, the business magazine *Caixin* reported that Jia had been linked to the corruption investigation of Ling Zhengce, a high-ranking Communist Party official from Shanxi, Jia's home province. Jia was friends with Ling's younger brother, Ling Wancheng, who was also arrested. The investigation, which turned into a wide-reaching scandal, would lead to the arrest of another brother, Ling Jihua, former president Hu Jintao's one-time chief of staff. Jihua was charged with taking bribes, obtaining state secrets, and abusing power. (He was jailed for life.)

On the face of it, the facts didn't look kind to Jia Yueting. His friend Ling Wancheng had established a private equity firm, Huijin Lifang, which, in 2008, became the second-largest shareholder in LeTV—behind only Jia himself. According to *Caixin*, registration documents revealed that one of Huijin Lifang's investors was a company called Beijing Jieweisen Technology, owned by a man named Jia Yunlong, a middle school teacher from Shanxi. Jia Yunlong told the magazine that he held the stake on behalf of his friend Jia Yueting, who had proposed using the teacher's identity to register the company.

The issues didn't stop there. When LeTV listed on the Shenzhen Stock Exchange in 2010, *Caixin* said, it received some special (but unspecified) assistance from the well-connected Ling Wancheng. Meanwhile, the investor's brother-in-law, Li Jun, had been LeTV's deputy general manager. Huijin Lifang ultimately sold its stake in 2011 for a profit of at least $45 million, according to conservative estimates, while Li Jun left in early 2013.

Jia has emphatically denied any corruption links. "The success of LeTV is based in no way on government connections," he told a Chinese news site. "I want no connection with politics." He said Huijin Lifang was a "normal investor" and did not provide any substantial benefit to the company. "Even though Huijin Lifang provided us financial support in the early stage of our company, if I were to make the decision again, I would not choose this kind of company as our shareholder."

Jia later explained his prolonged absence from China to Reuters, saying he was in California at the time to research Tesla and recruit a team for an electric car venture. "It was a difficult time for me because we faced a lot of external rumors and internal turmoil," he said. LeTV, trading under the name Leshi, was attempting to branch out beyond streaming, cell phones, and smart TVs, amid other expansion efforts, and if any one of them had failed, the company could have foundered. A Chinese news site reported that Jia had been seen in meetings in Silicon Valley, including with the chairman of Beijing Automotive Industry Corporation (BAIC). In August 2014, Jia had also said on Sina Weibo that he had visited Stuttgart, home to Mercedes-Benz and Porsche.

Jia had fair reason to worry about Leshi's financial complexities. In May 2016, *The Wall Street Journal* detailed his "unorthodox" methods of funding his operations. The newspaper noted that he pledged 85

percent of his shares in the company to lend to his new ventures—
including the smartphone and car businesses—in a personal capacity.
In 2015, Jia had sold more than $30 million worth of his company's
shares and then lent the profit back interest-free. Leshi explained the
practice to the *Journal* by saying Jia preferred to pledge shares rather
than bring in outside investors because he wanted Leshi to focus on
long-term growth. The risky maneuvers left the company vulnerable
to any sudden shifts in the stock market. For instance, any drop in
Leshi's share price could potentially trigger a margin call, which a
brokerage issues when it needs an investor to put up more cash to
meet the minimum requirement to keep holding a stock. In such an
event, Jia would have to sell his shares or find more funds to repay his
loan, jeopardizing his entire financing structure—and therefore every
company he touched—if his bets continued to go badly.

In early 2015, Leshi revealed its plans to combat China's air pollution, announcing that it would build an Internet-connected electric
car and that it already had 260 engineers in California. By August
that year, Jia was calling the planned vehicle LeSupercar and said six
hundred staff were already working on it. It later became apparent
that Jia was dedicating much of his energy to Faraday Future.

Back at 2016's CES, *TechCrunch*'s Burns turned the focus to what
Faraday's production model would look like. He quoted a journalist
who had seen the vehicle and described it as bigger, cooler, and more
futuristic than a Tesla Model X—"more daring and revolutionary
than anything from anyone else."

"So if it's really that good," Burns asked Sampson, "why show us
this thing?" He pointed at the picture of the car. "Why not show us
the good thing?"

Sampson seemed lost for a response. Burns continued for him.
"I'm *serious*. This was your big unveiling. You guys have been hyped
for *months*. It was like a blockbuster movie!"

Sampson held up his wrist to show that he was wearing a

black-strapped Apple Watch. "How many days before it was launched did you see one of these?"

"But this is from the most successful company in the world."

"And they don't show what they're doing until the day you walk through the store," said Sampson. (Not quite right—Apple typically reveals its new products several weeks before they're available in stores.)

"But you're an *unknown*," Burns insisted.

Sampson would not bend. "So, do we need to show anything?"

When it's speeding down a straight at 125 miles an hour and about to hurl itself around a sharp bend, a Formula E supercar sounds like a plane coming in to land. The car is electric, so there's no engine noise, but the tires make a racket on the tarmac and the slicing of the wind is fierce. Each car, made of a combination of aluminum, steel, carbon, and Kevlar honeycomb, carries a 440-pound battery and discharges its power so fast that a driver has to change steeds halfway through a race. After each race, the cars fuel up in tent garages, nurtured by generators powered by glycerin, a carbon-neutral by-product of bio-diesel production, in this case sold by a company called Aquafuel. It's not quite Formula 1, but it's not bad as a clean alternative.

Faraday Future was the title sponsor at the 2016 Long Beach ePrix, held in early April on a street circuit that was a slight variation on one used for the better-known Grand Prix of Long Beach, a high-octane event that attracts 200,000 speed junkies each year. It was the first motor racing event I had been to—save a dust-bowl demolition derby near the small city of Tauranga in New Zealand—so I spent my first hour wandering around the venue.

The city had fenced off a 1.25-mile loop of streets near the shore for the occasion. Grandstands, mostly empty when I arrived, had been set up trackside with large screens so fans could monitor the action from afar. Skyscrapers and hotels looked over the track from on high. Banners hung on aqua mesh affixed to wire fences and overpasses, proclaiming that visitors had jumped forward in time: THE FUTURE FORMULA E: THE NEW FULLY-ELECTRIC SINGLE-SEATER CHAMPION-SHIP. A parking lot next to the Long Beach Convention and Entertainment Center had been transformed into an "eVillage," with stalls manned by car companies showing off their electric models. Justin Bieber's music played loudly from speakers as the sun beat down from a cloudless sky and racing fans—mostly men—strolled aimlessly with beers in plastic cups.

The FFZERO1 was in attendance, too, parked in a booth by the entrance to the eVillage. A dozen or so Faraday staffers were there, wearing white T-shirts emblazoned with the FF logo—two uppercase *F*s linked and tilted forty-five degrees to the right to look like a stick-figure stealth fighter jet. Visitors crowded around for photos with the car, and early in the afternoon, Nick Sampson appeared, wearing a black baseball cap and dark sunglasses, to give reporters some boilerplate information about the would-be Batmobile.

But I hadn't come to see Faraday or Sampson. I was there to meet another Brit, Martin Leach, an auto industry veteran who had cofounded a company called NextEV. (Sadly, Leach passed away in late 2016.)

Like Faraday, NextEV, which later changed its name to Nio, saw the ePrix as a marketing opportunity. Like Faraday, Nio was backed by Chinese money (its early investors included the Lenovo Group, Joy Capital, Tencent, and Sequoia Capital China). Like Faraday, it had a presence in California. Unlike Faraday, it had a car in the race.

I met Leach by Nio's garage after the second practice race of the day. Nio's two cars were being fitted with new nose cones while they charged. A couple of improbably handsome young drivers with tousled hair and sculpted bodies had stripped off the top half of their driving suits to long undershirts. Leach, who was in his late fifties,

with whitening hair and a weatherworn complexion, looked somewhat less like a movie star. He wore a baseball cap and a crumpled suit
jacket paired with dark blue jeans. We shook hands and decided that
there was too much mechanical noise in the garage to have a produc-
tive conversation, so we walked across the tarmac, up the stairs of a
multilevel parking structure, and sat at a picnic table under a tent
where the teams would soon be eating a buffet lunch.

Leach, who spoke in a relaxed English accent, wanted to know
what my book was about. I told him it would be about the world's
transition to electric transport and its potential to activate a clean-
energy ecosystem. "Yeah, I think that's fair," Leach said, nodding.
"The car is such a fundamental part of everybody's life that when you
see electrification become significant in the industry, then that will
cause people to think about other things." The impact of electrification would be profound, he reasoned. "It's the most fundamental shift
in the industry, I would say, since Henry Ford invented the moving
production line."

Leach had some standing to make such a claim. He started racing
karts when he was eleven years old and turned professional at age
thirteen. Before he was eighteen, he had won the Europa Cup and had
come in third in the world championships. He was then about to start
his first full season racing single-seater Formula 4 cars, but his plans
changed after he was bedridden with rheumatoid arthritis. Being
from Essex, where Ford of Europe was headquartered, he decided
instead to ask Ford to sponsor his mechanical engineering degree at
Hatfield Polytechnic (now called the University of Hertfordshire). His
wish was granted. After graduating, he went to work for Ford and
returned to racing karts before again rising to the Formula 4 level. But
when he was twenty-six years old, the rheumatoid arthritis returned.
This time he had to choose between racing and his career at Ford. He
opted for the latter, a decision that would see him move from engineering into product planning and then into sales and marketing over

the span of four decades in the auto industry. During his career, he served as president of Ford Europe, director of global research and development at Mazda, and, for a year, chief executive of Maserati in a turnaround effort before it was split off from Ferrari and aligned with Alfa Romeo under Fiat.

In 2004, *Automobile* magazine named Leach its Man of the Year, even though he was jobless after leaving Ford Europe as a result of, in his words, being "drawn into the political ping-pong initiated by Ford worldwide." (Leach sued Ford in 2003 in a successful effort to prevent the company from enforcing a noncompete agreement.) "He's a certified car nut, one of the very few visionaries of the trade, an excellent engineer and driver, a pragmatic team player, and a genuinely nice guy," *Automobile* said.

In the background in Long Beach, a voice over the loudspeakers attempted to elicit cheers that were halfheartedly granted from a crowd in the eVillage. Hundreds more people got over their morning hangovers and started pouring in through the gates to catch the day's big races. Later in the afternoon, Leonardo DiCaprio, drinking champagne and wearing a gray flat cap, would make an appearance in the stands. DiCaprio, a Tesla owner, had been appointed chairman of the Formula E Sustainability Committee, established to promote the mass use of electric vehicles.

In the less glamorous surroundings of the parking lot lunch buffet, Leach told me the story of how Nio came into being. Leach was running his consultancy, an automotive services group called Magma International, when he was introduced to William Bin Li through a mutual acquaintance. Li was the CEO and founder of Chinese online automotive marketplace Bitauto, which has been described as like Cars.com, Autotrader, and *Consumer Reports* rolled into one. In 2016, the publicly traded company reached a valuation of about $2 billion.

Leach and Li first met in the coffee bar at the Paris Motor Show in October 2014. They discussed their concerns about the auto industry and their shared vision for what the future could be. Li wanted to start an electric car company with a user-centric model, but the Internet entrepreneur knew he needed a partner with an automotive background. Leach told Li about a 2011 experiment his consultancy had run for Fiat in Birmingham, England, called Fiat Click. The program allowed customers to browse products, configure a car, and then buy it online. If they wanted to see the car in the flesh, they could visit an Apple-like Fiat store in a high-end shopping mall. The sales experience was similar to a model being pioneered at that time by Tesla in the United States.

In 2009, China had surpassed the United States to become the world's largest passenger-car market, and its growth wasn't about to stop anytime soon. At the same time, the Chinese government was encouraging the development of "new energy" vehicles, and automakers in Western markets were facing increasing pressure to limit carbon emissions from their fleets. Like Faraday's founders, Leach noted that the convergence of electrification, connectivity, car sharing, and autonomous driving technology presented an opportunity to get into the game. "It is without a doubt the most exciting time to be in this industry," Leach told me as he sipped from a bottle of water.

Leach came away from his meeting with Li knowing that the pair would start a company together. "I felt I'd found somebody that had the same vision, the same alignment, and also could manage the wherewithal to make it happen, because it takes a lot of money and it takes patient investment," Leach said. "I've done over two hundred and fifty product programs and run five different OEMs [original equipment manufacturers]. I thought, 'I want to do that before I don't do anything else.'" (Leach stayed true to that wish. At the time of Leach's death, Li issued a statement to say that his legacy would stay with the company forever. "Martin is a true warrior and lived to the highest standard," Li said. "Even during his last minutes, he was still caring about the progress of NextEV.")

By March 2015, Leach had given up his other jobs and committed himself full-time to Nio. At that time, there were ten people at the start-up. It didn't take long for the company to start hiring aggressively, especially in China. By the end of 2015, it had close to eight hundred employees, and in December it announced a huge hire. Padmasree Warrior, former chief technology and strategy officer at Cisco, had signed on to be Nio's US chief executive and head of software development. After the announcement, Warrior, a Silicon Valley legend who had also served as Motorola's chief technology officer during its Razr flip-phone-driven heyday, told reporters that Nio would develop affordable electric cars for China before entering other markets. Leach would later tell me that the company was targeting a sub-$50,000 price for its first passenger car. The company revealed a concept of the car in March 2017. The Nio Eve, which looks like a stretched SUV, would be ready for the Chinese market in 2018. An updated version capable of full self-driving would be on sale in the United States in 2020, the company said. The car is designed for the autonomous era, with seats that can recline flat and tables that can unfold for use by the passengers.

A notable absence from the Long Beach ePrix was Tesla, which had resisted invitations to get involved with the event, but its presence was still felt. Nio and Faraday Future would likely not have been there were it not for Musk's company. "What Tesla has done and what Elon has done has influenced everybody," Leach said. "He's the only person who's successfully started up an automotive enterprise of any significance in decades." Leach also said that, in 2006, Musk had almost offered him the Tesla CEO job. It was during Tesla's early days, before the Roadster went on sale. "I had a conversation with Elon and he explained how he'd started PayPal and he had sold that, and now he was starting a car company, and he was going to do electric propulsion, and would I be interested in the CEO position." Leach was preoccupied with his other ventures—he had just been appointed chairman of the British van maker LDV—so he told Musk that he couldn't accept the job. "I wished him the best of luck and said it wasn't possible."

In contrast to Faraday, Nio had tried to tamp down hype surrounding the company, despite being at a similar stage of development and growing at a stupendous rate. By 2016, Leach's company was approaching a thousand employees in size and had signed an agreement with the Nanjing Municipal Government in China to spend half a billion dollars on a factory that would be capable of producing 280,000 power trains a year. While its headquarters would remain in Shanghai and its design studio in Munich, it also opened an office in San Jose, where it would add more than nine hundred workers by 2020 and invest $138 million in California, according to filings with the state's economic development agency, which granted the company a $10 million tax credit. By the end of 2016, it would pull back the covers on a million-dollar 1,360-horsepower supercar that it said would "outperform all combustion (engine) supercars in the world." In March 2017, Chinese Internet search giant Baidu announced that it, too, was investing in Nio. Then, in December, it lifted the covers on the Nio ES8, the update of the Eve, a seven-seater SUV with 220 miles of range and a starting price of US$83,000. But when I sat down with Leach, there had been no hyperbolic videos, no bold proclamations of beating Tesla at its own game, and no flashy stage shows. "We don't believe in talking about what we're going to do," Leach said. "We only want to talk about what we've done and what we've achieved."

He was echoing a comment that cofounder Li had earlier made in an interview with *Bloomberg*:

"Talking big is pointless."

There's a reason that Faraday Future and Nio are in California. Nowhere else on the planet can you find such ideal conditions for the germination of a new auto industry.

First, there are market considerations. California is the biggest
auto market in the United States and, thanks to tough emissions standards, the biggest market for alternative-fuel cars. The hybrid Toyota
Prius was the top-selling car in the environmentally conscious state in
2012 and 2013, before ultimately being surpassed by the Honda Civic
and Accord. It is Tesla's biggest market in the United States. The cities
of Los Angeles, San Francisco, and Sacramento were also among the
few in the country where General Motors' ill-fated EV1 was made
available in the 1990s.

Then there are the regulatory considerations. California has been
an international leader in establishing strict carbon emissions regulations. In 1990, the state enacted a rule that required 10 percent of vehicles for sale in California to be zero-emission vehicles (ZEV) by
2003. To accommodate hybrids, the California Air Resources Board
(CARB) later introduced credits for "partial zero emission" vehicles
and enabled a scheme that allowed automakers to trade credits to meet
their ZEV obligations—a feature that would become an important
source of income for Tesla. After legal challenges by automakers,
CARB adjusted its targets so that, in 2015, only 2.7 percent of new cars
sold within the state had to be zero emission. But it would gradually
increase the quota so that, by 2025, the figure would be closer to 22
percent.

Nine other states—including New York, New Jersey, Maryland,
and Massachusetts—ultimately adopted California's rules. By 2016,
these markets accounted for 28 percent of new vehicle registrations.
CARB's proponents, meanwhile, have figured that the impact has
been much greater. "Every car in the world is much cleaner burning
than cars before California's regulations were put in place beginning
in the 1960s," Dan Sperling, a CARB member and director of the
Institute of Transportation Studies at the University of California,
Davis, said in 2012. "All of them have emission control technology
that can really trace their history back to California." The automakers,

unwelcoming of the bureaucratic burden, were less cheery in their assessment. "We support a strong and comprehensive national policy that could erase administrative and logistical burdens," Mercedes-Benz USA said in a written statement in 2016. Translation: "We'd really prefer to avoid handicapping our gasoline-fed profit-makers on a state-by-state basis, if you don't mind."

Perhaps more important than the electric-car-friendly regulatory environment, however, is California's entrepreneurial mind-set.

On September 23, 1997, Steve Jobs walked onstage to applause at a meeting of Apple executives and managers. He had returned to the company two months earlier as interim CEO and was determined to make changes. Dressed in sandals, shorts, and a black mock turtleneck with which the world would later become familiar, he looked relaxed, even though he had been up until 3:00 A.M. working on an ad campaign that he hoped would revive Apple's brand. The brand had suffered from neglect since he left the company in 1985. Jobs didn't mention it in the meeting, but the brand wasn't all that was suffering at Apple. By 1997, the company had just 4 percent of the personal computer market and that year stood to lose more than a billion dollars. It was months away from bankruptcy. One of the first things Jobs wanted to do was bring public attention back to Apple's core values. "Our customers want to know, 'Who is Apple and what is it that we stand for?'" he said, pacing the stage. "Where do we fit in this world?"

To come up with the campaign, he had again teamed up with the advertising agency Chiat\Day, which had produced the famous Ridley Scott–directed "1984" ad to introduce the Apple Macintosh. "Apple at the core—its *core value*—is that we believe that people with passion can change the world for the better," Jobs said. He brought his hands together as if in prayer. "Those people that are crazy enough to think they can change the world are the ones who actually do." His speech seemed unpracticed and he used no notes. "The theme of the campaign is 'Think different.' It's honoring the people who think different and who move this world forward."

A video started on a screen at the side of the stage. The film opened with black-and-white footage of Albert Einstein smoking a pipe. "Here's to the crazy ones," said the voice of the actor Richard Dreyfuss drifting into the room. There was a quick cut to Bob Dylan, seen in profile. Piano and cello played in the background. "The misfits, the rebels, the troublemakers. The round pegs in the square holes." *Richard Branson*. "The ones who see things differently." *John Lennon*. "They're not fond of rules, and they have no respect for the status quo." *Thomas Edison*. "You can quote them, disagree with them." *Muhammad Ali*. "Glorify or vilify them." *Maria Callas*. "But the only thing you can't do is ignore them." *Gandhi*. "Because they change things." *Amelia Earhart*. "They push the human race forward." *Alfred Hitchcock*. "And while some may see them as the crazy ones . . ." *Jim Henson*. "We see genius." *Frank Lloyd Wright*. "Because the people who are crazy enough to think that they can change the world . . ." *Pablo Picasso*. "Are the ones who do." The film finished with a close-up of a young girl who opened her eyes and looked directly into the camera.

The ad signaled the start of an amazing recovery for Apple, the beginning of its transformation into the world's most valuable company. In the next twenty years, Apple would release the iMac, the iPod, the iPhone, and the iPad. Its market capitalization would vault from $3 billion to $700 billion. But the ad also expressed and perpetuated a conviction that Californians—and Silicon Valley in particular—could change the world.

It's easy to ridicule Silicon Valley's "change the world" pretensions, and many have made sport of doing so. HBO's *Silicon Valley*, for instance, has been one of the most effective ("I don't want to live in a world where someone else is making the world a better place better than we are," says a chakra-channeling CEO in the show's second season). But while unselfconsciously dweebish, the "change the world" mentality is not entirely delusional, or is, at minimum, a useful article of faith. Silicon Valley has produced numerous world-changing products and services. The semiconductor. The personal computer. The graphical user interface. The Web browser. Google Search. The touch screen smartphone. Social networks. All of these things, if not created in Silicon Valley, were at least perfected there. The same will likely be true for virtual reality, artificial intelligence, and self-driving cars—all of which stand to have seismic effects on the way humans live.

Essential to the drumbeat of innovation in Silicon Valley is the idea that one can "make a little dent in the universe," as Jobs said in a 1985 interview with *Playboy* magazine. Even if these beliefs are in many cases mistaken, they at least help build a supply chain of brains that makes Silicon Valley unusual. Fed on a diet of inspirational quotes from their tech heroes ("Remembering that you are going to die," said Jobs, "is the best way I know to avoid the trap of thinking you have something to lose"), the best developers and software-minded entrepreneurs stream into the Valley to make an impact, get a view of history in the making, make a chunk of money, or all of the above. And it works.

Silicon Valley's mind-set is one of embracing risk and not being afraid of failure ("In a world that's changing really quickly," said Mark Zuckerberg, "the only strategy that is guaranteed to fail is not taking risks"). Young acolytes are inculcated with these teachings, committing them to memory so they can recite them on demand. And then, for better or worse, they act on them. A 2012 Harvard Business School study estimated that 75 percent of start-ups fail—that is, they run out of money or fall to pieces before becoming profitable. But that's okay in Silicon Valley. In fact, it's encouraged. "If things are not failing, you are not innovating enough," Musk has said. Like Hollywood, it's a hit-driven business. A venture capitalist will place bets on a number of promising start-ups with the expectation that most will come to nothing but with the hope that one will become a Google or a Facebook. Better to have failed trying than to have failed to try.

Even the Valley's most humiliating moment of failure—the 2000 dot-com crash—has ultimately had some positive side effects. You could say, in fact, that the dot-com bubble helped get electric cars to where they are today. While the period from 1997 to 2000 in the Valley has served as a painful lesson in irrational exuberance, it did at least provide a setting for a young Elon Musk to sell his mapping start-up, Zip2, to Compaq in February 1999—close to the height of the bubble—for about $300 million, helping him fund the start-up that would become PayPal. As previously mentioned, Musk's proceeds from the sale of PayPal in turn helped him start SpaceX and Tesla.

Tesla's rise has catalyzed the development of a new-auto ecosystem in California. Since 2010, when it took over the shuttered NUMMI factory in Fremont, suppliers have come to the region, or expanded their existing operations, to meet Tesla's needs.

A drive around the business parks in and around Fremont takes not much more than half an hour. I did the drive in my wife's Honda Civic on a hot, sunny day in June 2016 to get a look at some of Tesla's suppliers. In neighboring Newark, I saw the North American headquarters for Australian company Futuris, which had taken over a 160,000-square-foot former Staples distribution center for its car-seat-making operations. The company was once based in Tesla's Fremont factory, but it outgrew the space. Its new facility sat at the end of a cul-de-sac just beyond freshly developed business parks with FOR LEASE signs posted on boards outside. (Since Tesla decided in 2015 to manufacture its seats in-house, its relationship with Futuris appears to have been scaled back.)

Driving farther south alongside the bay, with dry-grass hills to the east, I next stopped by Toyota Tsusho America (TAI), a metals and electronics supplier that occupies one of a series of vast beige cuboids by a two-lane expressway. In 2014, it had signed a five-year lease for the warehouse space after it, too, had been bumped out of the Tesla factory. A block away, a road construction crew was laying asphalt outside a new business park.

Closer to the shore, near the former Fremont Dragstrip, was Asteelflash, which made circuit boards for Tesla's vehicles. In recent years, it had spent several million dollars adding production capacity and an automated coating machine to serve Tesla's needs as it grew. Across the highway, I could see the tops of the Tesla factory's pearl-white walls. The chrome letters T-E-S-L-A, each one itself the size of a building, were written on the side and visible from miles away.

After my self-guided tour of Fremont's industrial wonders, I ate lunch at a packed Thai restaurant downtown. It was opposite a construction site and a sign that read: FREMONT DOWNTOWN ON THE RISE. Fremont was in the midst of transforming itself from commuter suburb to high-tech manufacturing hub. Five miles away, behind the Tesla factory, was the newly built South Fremont station for the Bay Area Rapid Transit trains that crisscross the region. Next to that was South Fremont's Warm Springs Innovation District, which was in the process of converting 850 acres of land into a housing, shopping, and entertainment hub, with hotels, convention facilities, and parks. The land had previously been zoned for heavy industry. Six years after NUMMI closed, taking close to five thousand jobs with it, Fremont was on the rebound.

Tesla is also partly responsible for the emergence of an automotive ecosystem on the other side of the bay, in Silicon Valley. The companies there are almost exclusively focused on software.

"For a hundred years, automobiles have been a mechanical engineering industry. Now, there is the shift to software—and the mecca of software is Silicon Valley," Dragos Maciuca, director of Ford's Palo

Alto research and innovation center, told the *Los Angeles Times* in late 2015. Ford, Toyota, Honda, Hyundai, Volkswagen, BMW, Mercedes-Benz, GM, Nissan—they've all established Silicon Valley–based research centers to work on autonomous driving and connectivity. Automotive suppliers Continental, Delphi, and Denso also have offices in the area. Local tech companies are branching out into automotive, too. Santa Clara–based graphics chipmaker Nvidia has added hundreds of engineers to its auto-focused teams in the past few years. "We didn't start out to be an auto company," Danny Shapiro, Nvidia's senior director of automotive, told the *Times*. "But everything that is changing a car has nothing to do with the auto industry of the past."

Start-ups have spotted the opportunity, too, of course. Uber and Lyft, both based in San Francisco, are hogging the early spoils in the ride-sharing market. Younger companies like Mountain View's Smartcar (infrastructure for the connected car), San Francisco's Reviver (digital license plates), and Palo Alto's Nauto (AI-powered autonomous driving) are pursuing other software-related opportunities. Meanwhile, electric power-train companies like Wrightspeed (heavy-duty trucks), Zero (motorcycles), and Proterra (buses) are also in the area and have collectively raised hundreds of millions of dollars in funding.

On the autonomous-driving side of things, Alphabet (formerly Google), which has logged several million self-driving-car test miles, continues to lead the pack. At the end of 2016, it created a new business division, called Waymo, for its autonomous driving technology. In May 2017, Waymo and Lyft announced that they would work together on developing the technology, and later in the year, Alphabet invested $1 billion in the start-up. Others, like Cruise Automation (which GM acquired for $1 billion) and Comma.ai, which offers open-source autonomous driving technology in the same vein as Google's Android mobile operating system, are chasing hard. Baidu, China's leading Internet search company, has an autonomous-driving research center in Sunnyvale. Byton—backed by China's Tencent, Foxconn, and the China Harmony New Energy auto retailer group—has an office in Mountain View, as does Didi Chuxing, the Chinese ride-sharing company in which Apple invested $1 billion.

Many of these companies have taken not just inspiration but also talent from Tesla. Part of the value of an innovation cluster like Silicon Valley lies in the dispersal of intellectual labor from one node to the next. For instance, PayPal is well known in the Valley for producing a number of high performers who left the company to start, join, or invest in others. The so-called PayPal Mafia includes Reid Hoffman, who founded LinkedIn; Max Levchin, whose most recent of several start-ups is the financial services company Affirm; Peter Thiel, a Facebook board member and President Trump–supporting venture capitalist who cofounded "big data" company Palantir; Jeremy Stoppelman, who started reviews site Yelp; Keith Rabois, who was chief operating officer at Square and then joined Khosla Ventures; David Sacks, who sold Yammer to Microsoft for $1.2 billion and later became CEO at Zenefits; Jawed Karim, who cofounded YouTube; and one Elon Musk.

In a similar way, Tesla alumni are involved with many of the ventures that make up Silicon Valley's new-auto ecosystem. Ian Wright, the founder of Wrightspeed, was one of Tesla's five founders and left the company in 2005. Proterra's CEO is Ryan Popple, a former senior finance director at Tesla (he left in 2010). Cruise recruited Andrew Gray from Tesla's Autopilot team. Self-driving car start-up Comma .ai snared former Tesla senior system engineer Riccardo Biasini. Former Tesla Autopilot head Sterling Anderson left to start autonomous driving company Aurora (which has since announced a partnership with Byton). And Byton has nabbed a handful of former Tesla highfliers, including erstwhile senior manager of vehicle engineering Paul Thomas, former director of supply-chain manufacturing Mark Duchesne, and former director of vehicle purchasing Stephen Ivsan.

Then there's Apple.

"They have hired people we've fired," Musk told the German newspaper *Handelsblatt* in September 2015. Rumors that Apple was working on an electric car project had emerged that February, when an Apple employee allegedly e-mailed *Business Insider* to say that the Cupertino company was working on a project that would "give Tesla a run for its money." Former Tesla people had been joining the company in droves. Musk had his own view of the situation. "We always jokingly call Apple the 'Tesla Graveyard,'" he said. "If you don't make it at Tesla, you go work at Apple. I'm not kidding."

Steve Jobs had floated the idea of making a car in several discussions in 2008, according to former Apple vice president Tony Fadell, who went on to start the smart-devices company Nest, which was eventually acquired by Google (and which he left in 2016). Jobs and Fadell had discussed the proposition during a few walks, Fadell told an interviewer in 2015. "A car has batteries; it has a computer; it has a motor; and it has mechanical structure. If you look at an iPhone, it has all the same things," Fadell said. "But the hard stuff is really on the connectivity and how cars could be self-driving." In fact, Apple had even considered building a car before it released the iPhone in 2007, according to court testimony from Phil Schiller, Apple's senior vice president of marketing, in 2012.

In 2015, with more than $200 billion cash in hand and a perceived need to expand beyond the slowing smartphone market, an Apple car project suddenly started to seem more feasible. After *Business Insider*'s scoop, reports emerged that the company was poaching people from the automotive industry and planned to hire more than a thousand people for the so-called Project Titan. Among the most prominent hires was Chris Porritt, Tesla's former head of vehicle engineering, who had joined Musk's company from Aston Martin, where he had been chief engineer on the One-77 supercar project. When news broke of Porritt's hire, Tesla noted that Porritt had been out of the company for seven months.

Apple, which is one of the most secretive companies on the planet, has not revealed any details of its car plans, so the press has been left to do its reporting based on speculation or leaks. In Sunnyvale, neighbors of a fenced-off Apple-occupied facility reported mysterious noises—bangs, thumps, beeps, whines, and hums—that would come in the middle of the night. Both BMW and Daimler had broken off talks with Apple about working together on a car, a German newspaper reported. Apple had been meeting with charging-station companies with a view to laying the groundwork for a charging infrastructure for a self-driving electric car, said another report. In April 2016, grasping for something to write about, *Motor Trend* hired a graduate student to mock up a design of what an Apple car could look like and then convened a panel to spitball ideas of how such a product might work. "The glazing would be beautiful, well-proportioned with some automotive cues that look sure-footed and capable, not cutesy," one of the panelists imagined. "Approaching it will be like walking up to an amazing store in Tokyo; the way the door opens up and presents isn't a door you grab but a roof that raises and you walk in."

Later in the year, Apple appeared to have scaled back its electric car ambitions, shedding hundreds of members from the Project Titan team, according to reports. Bob Mansfield, an Apple veteran who had overseen the iPad, was appointed to lead the project and had refocused its development efforts on an autonomous driving system. The approach would allow Apple to keep its options open on the question of whether to produce its own car.

There's one more Silicon Valley car company that is part of the Tesla diaspora. Intriguingly, it has something in common with Faraday Future: an investor by the name of Jia Yueting.

Lucid Motors was started under the name Atieva (which stood for "advanced technologies in electric vehicle applications" and was pronounced "ah-tee-va") in Mountain View in 2008 (or December 31, 2007, to be precise) by Bernard Tse, who was a vice president at Tesla before it launched the Roadster.

Hong Kong–born Tse had studied engineering at the University of Illinois, where he met his wife, Grace. In the early 1980s, the couple had started a computer manufacturing company called Wyse, which at its peak in the early 1990s registered sales of more than $480 million a year. Tse joined Tesla's board of directors in 2003 at the request of his close friend Martin Eberhard, the company's original CEO, who sought Tse's expertise in engineering, manufacturing, and supply chain. Tse would eventually step off the board to lead a division called the Tesla Energy Group. The group planned to make electric power trains for other manufacturers, who needed them for their electric car programs.

Tse, who didn't respond to my requests to be interviewed, left Tesla around the time of Eberhard's departure and decided to start Atieva, his own electric car company. Atieva's plan was to start by focusing on the power train, with the aim of eventually producing a car. The company pitched itself to investors as a power train supplier and won deals to power some city buses in China, through which it could further develop and improve its technology. Within a few years, the company had raised about $40 million, much of it from the Silicon Valley–based venture capital firm Venrock, and employed thirty people, mostly power train engineers, in the United States, as well as the same number of factory workers in Asia.

By 2014, it was ready to start work on a sedan, which it planned to sell in the United States and China. That year, it raised about $200 million from Chinese investors, according to sources close to the company. Beijing Auto (BAIC), through its subsidiary Beijing Electric Vehicle Company, committed $100 million in May 2014 and was joined a couple of months later by Jia Yueting's Leshi, which contributed the lion's share of the remaining $100 million.

In the wake of the funding, Atieva attracted some major talent, including—in a nice piece of circumnavigation—Martin Eberhard, and Derek Jenkins, a star designer who had risen through the auto industry's ranks at Audi, Volkswagen, and Mazda alongside Tesla chief designer Franz von Holzhausen. Jenkins designed the 2016 Mazda MX-5 Miata and the Mazda CX-9, among other highly regarded Mazda models.

But in the two years following BAIC's 2014 investment, Atieva's relationship with the Chinese automaker, which owned about 25 percent of the company, deteriorated. BAIC wanted Atieva to build a car only for China, but Tse and company wanted to build first for the United States to establish the company's brand, then to enter China one or two years later. BAIC bristled at this approach and convinced the rest of the board to support its China-first plan. Fed up with the political battles, Tse quit. Eberhard, who had joined soon after BAIC invested, had already beaten him out the door, complaining it was being run like "an old-school Hong Kong company." He had lasted six weeks in the job. (In 2017, Eberhard became involved with SF Motors, a US subsidiary of a Chinese car company.)

Even after the departure of several other executives, however, the company continued to resist BAIC's will. BAIC eventually became so frustrated that, in March 2016, it sold its stake in a public auction in Hong Kong through its Beijing Electric Vehicle Company. (To be precise, the sale was made by Beijing Electric Vehicle Company's Hong Kong subsidiary, which, according to public business records, appears to have existed solely for overseas investments, had zero employees, and whose named legal representative was Xu Heyi, the CEO of BAIC.)

The buyer, according to public documents filed in Hong Kong, was a company called Blitz Technology Hong Kong, whose director was listed as Yi Hao. Yi has several ties to Leshi. For example, Yi had previously founded two companies with Leshi Investment Management, one of which was an e-commerce app called Jiuai (Old Love). Included among that company's investors were Leshi and the diamond retailer Meikelamei. Yi, who must be a busy man, was CEO of both companies that he cofounded with Leshi Investment Management, as well as CEO of Meikelamei and director of Blitz Technology Hong Kong.

Yi and a close associate of Jia Yueting were also among the biggest beneficiaries from the 2014 sale of Meikelamei to a company called Haoningda Meters Company. Yi owned 11.2 percent of Meikelamei. Another major Meikelamei shareholder was Beijing Guangmo Investment Company, the new name of a group previously known as Beijing Jieweisen Technology. That's the investment group that previously held a stake in LeTV through Huijin Lifang, the private equity firm caught in the corruption scandal that ensnared the high-ranking Communist Party officials from Shanxi. The man listed at the head of Beijing Guangmo was Jia Yunlong, the teacher who told *Caixin* that Jia Yueting had used his identity to register the company.

It would appear, then, that on top of Leshi's initial investment in Atieva, which gave it about 20 percent of the company, the financially compromised Jia Yueting was at the very least closely associated with the mysterious new company that had purchased BAIC's 25 percent share of Atieva in March 2016. Atieva thus became disproportionately reliant on funding from a source that had worryingly complex financial arrangements. The need to find new funding sources may have contributed to the sudden sense of urgency Atieva displayed when it finally came out of stealth mode in June 2016. It revealed to Reuters that it planned to start selling a premium electric sedan in 2018, followed by two luxury crossover utility vehicles in 2020 and 2021. The company was testing its dual-motor electric power train on a Mercedes-Benz Vito van, which could accelerate from zero to sixty miles per hour in 3.1 seconds (and, later, 2.69 seconds)—astonishing for a vehicle of that size.

The company also revealed that its funding totaled several hundred million dollars. Designer Derek Jenkins said that it would set itself apart from an emerging field of rivals in China with its "California mind-set."

Four months later, Atieva had changed its name to Lucid Motors. In December, it revealed a prototype of the car, called the Air, to the public and said one version would come with a 130 kilowatt-hour battery pack and drive four hundred miles per charge. The car had a motor on each axle, and the company claimed it would draw on 1,000 horsepower, helping it launch from zero to sixty miles per hour in 2.5 seconds. It sported a full glass roof and four lidar sensors for autonomous driving, and the company had plans for a luxury interior with two overstuffed rear seats mimicking the feel of first-class air travel. The exterior was the size of a midsize car, like the BMW 5 Series, but had the roominess of a large sedan, like the BMW 7 Series—a feat made possible by the electric power train, which takes up less space than an internal combustion engine setup. The sleek sedan would ultimately sport interior design themes inspired by California: early morning in Santa Monica, midday in Santa Cruz, and midnight in the Mojave Desert. To maintain quality, the company planned to produce eight thousand to ten thousand cars at a factory in Arizona in its first year, starting late 2018. It would later ramp up to fifty thousand to sixty thousand cars a year. The base version of the Air would start at $60,000, with more luxuriously appointed versions—including larger battery packs, all-wheel drive, and a glass canopy roof—on sale for more than $100,000.

Lucid's moments in the spotlight also elicited one more intriguing detail. Its chief technology officer, and de facto CEO, was Peter Rawlinson, who had quietly joined the company in 2013. The British engineer would have been familiar to anyone who had followed Tesla's story closely over the past decade. He had been the original leader for development of the Model S. A former engineer at Jaguar and Lotus, Rawlinson ended up leaving Tesla at the same time as fellow Brit Nick Sampson, who had led chassis engineering.

Rawlinson's and Sampson's departures in January 2012 came just months before the Model S was due to go on sale. Tesla attributed Rawlinson's departure to a need to "tend to personal matters in the UK." Rawlinson didn't respond to my repeated requests for an interview, but Sampson would tell me that his former colleague went home for a vacation and didn't return. Tesla provided even less detail about Sampson's exit, saying only that he had "fully transitioned" off the Model S project when he left the company. Sampson joked that, in the wake of Rawlinson's unexplained absence, Musk had decided he hated the English, and so Sampson was "asked to retire."

Tesla's stock dropped 19 percent on the news of the engineers' departures.

—————～～～～～～～—————

I'm crouched as low as I can get to look at the underside of the front bumper on the DF 91, the project name for Faraday Future's first car (later named FF 91). After careful inspection, I determine that the bumper is black and shiny. I stand up and step to the side to peer at its front wheel arches. They are as lusciously curved as a marshmallow's behind. I step back to consider the machine from a wider angle. As far as SUVs go, this thing is beautiful, like a machine-gun bullet pregnant with twins. It has the kind of sculpted sci-fi artistry that would draw a crowd in a supermarket parking lot. The car sits on a rotating pad in the middle of a minimalistic showroom with space-gray walls. I look up and see the ragged peaks of an arid mountain range through a skylight. Then I look down at my hands. Wait. I have no hands.

Where have my hands gone?

I suddenly remember that I'm in an office in Gardena and wearing a virtual reality headset. The very real-looking car in front of me is not actually there at all. It's all computery make-believe intended to help Faraday work through design considerations. My hands are absent because my digits have not been digitally rendered for this virtual world. There are no mountains, there is no skylight. The carpet, as I'm reminded when I remove the headset, is ugly.

But *damn,* that was cool.

The virtual reality experience was a serendipitous moment. I was being shown around Faraday's headquarters by a young man from the public relations team, when we walked into a room dedicated to user interface design. First, I perused a wall of design concepts that demonstrated some remarkable but highly theoretical ideas: a Roomba-like robot that attaches itself to the underside of the DF 91 to charge the car; text that glows on the side of the vehicle when approached, greeting passengers by name; and an augmented reality overlay on the inside of the car's windows, so they are transformed into interactive displays. When I turned around, I noticed a huge computer under a desk in the corner of the room and the HTC Vive headset on top of it. I barely had to ask.

Faraday's offices are in the former headquarters of Nissan USA, which vacated the premises in 2006 when, after forty-six years at the site, it moved its operations and 1,300 jobs to central Tennessee. The main building is a long yellow block that, in profile, bears resemblance to an Edwardian chest of drawers. In 1960, it would have been the height of modernity. Across the parking lot, there's a misshapen annex that houses a reception desk and the executives' offices.

There are two wings on the second floor of Faraday's main building that accommodate the majority of the company's engineers, designers, and operations folks, and there is little regard for organizational aesthetics. Bodies are crammed in wherever space allows, and people are seated shoulder to shoulder at row after row of desks. There is a

general, pervasive sense of busyness. T-shirted designers lean over the
desks of their colleagues. An ad hoc meeting takes place by a monitor
in the corner. Someone laughs. Someone else calls out a nickname.
"Not doing any work today?" a passing colleague jokes to the guy
who's showing me around. Conference rooms line the flanks, and
they are mostly full. Inside each one is a large-screen LeTV monitor,
reminding occupants of the source of the company's funding.

Downstairs, there's a cafeteria populated with short tables and
plastic chairs. A blackboard stretches the length of one of the walls.
On it, an artist has drawn a chalk mural depicting the characters and
cars from the *Fast & Furious* movie franchise. The FFZERO1 had apparently been lined up to appear in the eighth *Furious* film, but the
plan fell through for reasons my host won't divulge. Outside, a few
more tables are shaded by colorful umbrellas.

Still downstairs, we walk through a shiny-floored workshop where
engineers tinker with electric motors, bundle up wiring harnesses, and
test battery cells and inverters in secure glass-walled boxes that look
like they're built to withstand a nuclear explosion. Minivans and
SUVs are kitted out with sensors and cameras to assess Faraday's
assisted-driving systems. Across a hallway, a heavy gray door opens
into a room where designers chisel away at half-painted clay models of
the DF 91. One such model sits under a cloth cover. Signs advise that
there are strictly no photos allowed in here.

Faraday's founding team had deep connections to Tesla, with
which the company shares some cultural similarities—as well as some
carefully cultivated differences. Nick Sampson, who got the company
running with his former Lotus colleague Tony Nie, was quick to recruit some pals from his Tesla days. Joining him in the leadership
ranks were human resources chief Alan Cherry, manufacturing head
Dag Reckhorn, and supply chain chief Tom Wessner, all of whom had
spent time at Musk's company. When Faraday made its first big press

announcement, in July 2015, it also revealed that it had hired several key engineers and designers who had worked at Tesla.

By the time I stepped into Sampson's office, it had been six months since he was scorched onstage at CES by the *TechCrunch* reporter. Sampson sat at his desk in a blue long-sleeve shirt with an orange lanyard around his neck. Through his office window, I could see a food truck in the parking lot. He had gray stubble and his hair was cropped close to the sides of his head, the top of which was bald.

Sampson had joined Tesla in 2009 after being approached by a recruiter through LinkedIn. By coincidence, he found that his former Jaguar and Lotus colleague Peter Rawlinson was already at the company. After Rawlinson convinced him that it was a good place to be, Sampson met Musk. At the time, Tesla's core engineers and designers were working in an assigned area within the SpaceX headquarters in Los Angeles. Sampson, who has an appealing inability to resist smiling while he speaks, couldn't recall anything specific about his first meeting with Musk, but he remembered that his soon-to-be boss was charismatic, had a clear vision, and was driven to succeed. He was attracted to Musk's plan to shred the normal way that car companies operated.

The Model S was in no shape to speak of when Sampson arrived at Tesla, even though the company had unveiled the show car, based on a Mercedes CLS, at the press event in March that year. "It was just a CLS with a Roadster motor and batteries stuffed wherever they could put them," Sampson said with a laugh. Franz von Holzhausen, the chief designer, had come up with the styling for the car, but there was not much done in the way of engineering. At the time, Sampson shared a desk in Tesla's section in SpaceX with manufacturing boss Reckhorn and supply chain guru Wessner. He smiled at the memories. It was a dynamic environment, he said. "I don't think you can beat that."

Sampson has sought to replicate that environment at Faraday, with supply chain people seated next to manufacturing folks alongside designers and engineers and marketers and HR and . . . You get the picture. Traditional automotive companies typically don't work like that. They like their lines of division.

Sampson left Tesla with mixed feelings. He was 100 percent behind Tesla's mission and thought the company's strategy was spot-on. But he had reservations about other areas. "It was not a positive culture," he said of his time there, the near-constant smile slipping momentarily from his face. He declined to go into specifics about what exactly he meant, but he attempted to explain it by metaphor.

"The guy who beats the world hundred-meter record doesn't do it because his coach is shouting at him or doing something to make him run faster," he said. "At the end of the day, that guy breaks the record because he wants to do it." A good coach can improve the runner's technique, Sampson suggested, and help him find the self-motivation that will spur him to put in that extra fraction of a percent that takes a tenth of a second off his running time. That's the approach Sampson was attempting to instill at Faraday. He gave the distinct impression that it wasn't present at Tesla. "We're trying to enable people, encourage people to push themselves rather than be pushed. You'll always push yourself far harder than anyone else can push you."

After leaving Tesla, Sampson got in touch with his former colleague Tony Nie, who had started Lotus Engineering in China and had since established a consultancy for new-energy vehicles. Tesla had opened its first store in Beijing, and there was some hype building around the company there. Tesla had made people fundamentally rethink electric vehicles, Sampson said, and proved that there was a way for start-ups to get into the auto industry. Sampson and Nie thought there might be a way to take advantage of China's sudden interest in electric cars and the country's governmental support for them. Others were thinking the same. The two men were approached by numerous China-based groups that were interested in starting a car company, including a consortium of parts suppliers and a drinks maker, but there was one man who stood out: Jia Yueting wanted to know what it would take to start a company that could be better than Tesla.

Sampson first encountered Jia when he traveled to China to witness the launch of some of LeEco's new devices in April 2014. He had done some online research about Jia, but didn't find out much. Jia was still a largely unknown character, even in China. Sampson showed up to the event in Beijing expecting to find a besuited CEO behind a lectern in a lecture hall with a shoddy PowerPoint presentation. Instead, he stepped into a massive auditorium filled with people. Jia came out in jeans and a black T-shirt with giant screens behind him. "Oh, it's a Chinese Steve Jobs here!" Sampson thought. Jia spoke for an hour in Mandarin, and even though Sampson couldn't understand a word, he could tell that the audience was rapt. "Just even in his initial style—it was Californian, not Beijing," said Sampson.

Sampson and Nie spent several days with Jia in Beijing as the entrepreneur explained his vision for bringing a "digital lifestyle" to cars.

These conversations formed the basis for what would become Faraday's business model. The company would start by selling premium cars in China and the United States but design them from the inside out, with the customer experience top of mind. The cars would be built to be fully autonomous so, when the technology and regulations allowed, passengers could relax and enjoy the ride on reclining seats surrounded by touch screens that would stream LeEco's video content. One day, the cars could be sold cheaply—perhaps even below the cost of making them—and bundled into a subscription package that would include other LeEco services (LeEco was pursuing such a model with its TVs and smartphones). Alternatively, the cars could be available through an Uber-like car-sharing service, meaning no ownership was required. And of course, the cars would be electric. At one point during the meetings, Jia looked out the window of his sixteenth-floor office and said, "I want to do it in a clean way. I don't want to just put my technology into gasoline cars."

After meeting Jia, Sampson knew immediately that he had found a partner—not least because Jia agreed, without hesitation, that the company had to be built in California for it to be a global success. Sampson also noted that Jia had some commonalities with Musk. Both men were driven, passionate, and totally committed to their missions.

Like Musk, too, Jia had his own way of thinking about innovation.

To make great advances, he believed, you should not look to the future and try to figure out how to get there. Instead, you should start *from* the future and look back on how you traveled there. This was a view that resonated with statements from Musk, who once said: "The first step is to establish that something is possible, then probability will occur."

As if to prove his change-the-world credentials, the Chinese entrepreneur even had his own Silicon Valley-esque mantra to abide by: "You have to imagine the future from the future's perspective."

No doubt about it. Jia Yueting was one of the crazy ones.

9

BUILD YOUR DREAMS

"What does the real Chinese customer need?"

Lei Ding, head of LeEco's auto division, stood onstage in a hall large enough to host a U2 concert. It was April 20, 2016, and he was dressed in a black blazer and designer spectacles. Behind him was a screen the size of half a basketball court, and in front of him was a long catwalk flanked by rows of men wielding smartphones with screens that flickered like glowworms in a cave. He clicked through slides with a handheld remote. Earlier in the day, LeEco had unveiled its latest range of smartphones and smart TVs to ten thousand fans, media, and other guests. But now, at a Beijing sports arena that the company itself owned, it was time for the surprise main event.

Lei showed a picture of the LeSee, a concept car to build hype for LeEco's vision of the auto future. It shimmered gloriously in white on the screen behind him, a sedan fit for James Bond's family, with a glass dome like half a volleyball sitting on the roof. "It's a combination of tradition and modernity," Lei declared. Seated in the audience, young women in evening dresses and men in tuxedos were mixed in with millennials in T-shirts and office workers in business casual. Lei directed their attention to the car's modern lines and glass roof. On its front, in place of headlights, the car wore a thin LED smile. A 3-D rendering showed a postmodern interior with a steering wheel that folded into a recess when the car was in autonomous mode, and a back seat of terraced white foam that would shape itself around a passenger's body. Lei called up images that showed LCD touch screens in front of the passenger seats. Each passenger would have isolated audio so what they were listening to or watching wouldn't affect others in the car. It was a rolling personal movie theater.

"But language is very weak compared to action," Lei said. It was time to show a video of a hypothetical car in hypothetical action. A series of photorealistic animations danced across the screen, backed by an electro soundtrack that mixed industrial synth beats with a ruminative string chorus. The camera panned luxuriantly over design details—strakes and beltlines, a hubcap in close-up, an illuminated LE logo on the steering wheel—and then showed the self-driving vehicle speeding along an empty road in a sci-fi cityscape. The video finished with the car pulling up to a *Star Trek*–esque spaceport—or maybe it was just an empty airport from the 2050s—to meet a man in a black long-sleeve T-shirt. The audience responded with wowed applause.

Jia Yueting stepped out with a giant smile. He was dressed like the man in the video and in his right hand he held a plus-size LeEco smartphone. The LeSee, he announced, carried the company's dream of improving people's lives and the environment. "I'm sure you're all very eager to see it in person, right?" There was a cheer and Jia's smile broadened. "I don't think I hear enough enthusiasm!" The crowd added more gusto to its hollers. "All right," said Jia. "I'm sure I will not let you down."

Such techno-rapturous scenes are still relatively new to China.

While Internet companies like Alibaba and Tencent have been staging large-scale company events with elaborate karaoke and dance performances in sports stadiums for more than a decade, local tech companies came late to the art of the product launch. And it looks like they learned it from Apple. Jia applied a Jobsian rubric to selling smartphones: He paced the stage while fetishizing his company's products and their apparently magical qualities.

"How do we get this car up here?" he asked the crowd. He looked at his phone, which he was brandishing, trophy-like, near his head. "I'm going to call our car out with this phone." He spoke into the device, and a map flashed up on the screen, identifying the car's location. After a few seconds, Jia received a message. He told the crowd what it said: "A driver named Ding has accepted my order!"

Ponderous drums started thumping over an orchestral arrangement of titanic quality. Jia turned his attention to a shipping container at the end of a catwalk. The container's doors opened and clouds of dry-ice fog spilled out. Then, creeping in from the shadows, came the LeSee, heralded like a chariot of the gods. Lei Ding was behind the wheel as the car rolled forward at approximately three miles an hour. Colored lights strobed in the background. "Whoa!" Jia cried in excitement. "*Whoa!*" The car parked itself on a rotating platform and was spun so it could be ogled from all angles. (Eight months later, *Buzz-Feed* would reveal that the performance was stagecraft—the car was being controlled from backstage by remote control. But the crowd didn't know that.) People sprung to their feet with smartphones held high to photograph the gleaming machine.

"This is such an emotional moment for me," Jia revealed. "When everyone is questioning us over our ability to develop a car like this, and is laughing at us"—he didn't mention who—"we are still able to be here and show you this car."

Jia and Lei posed for photos with their baby before concluding the ceremony. When it was time for the vehicle to return to its box, Jia again spoke into his smartphone. As the LeSee crept forward without a human driver, he raised his right fist and gently punched the sky.

———∿∿∿∿∿———

You never have to look far to find scenes of change in China, but the sense of dynamism is perhaps nowhere more profound than in the border city of Shenzhen. In the 1970s, Shenzhen was an unremarkable fishing village at the end of the Kowloon-Canton rail route. Since President Deng Xiaoping established it as a Special Economic Zone in 1980 as part of the opening up of China's economy, it has been on a mercantile tear, its population exploding to twelve million people. Today, Shenzhen is a booming metropolis, overflowing with energy and optimism. It is a beacon for young people who want to get ahead in business or score a job at one of the city's tech companies, like electronics manufacturer Huawei, Internet giant Tencent, or the iPhone-producing Foxconn. Migrants from other parts of China make up more than 80 percent of the city's population.

Shenzhen stretches its arms thirty-five miles wide as it hugs Hong Kong's New Territories. To drive from one side to the other is to pass bright new skyscrapers, shopping malls, convention centers, entertainment venues, and sports arenas of geometric radicalism, each competing to attain new levels of gobsmackery, preening under the weight of reflective domes, fine latticework, structural cowlicks, honeycombs, razor edges, suspended ledges . . . It's not a mere excitement of the architectural senses—it's a raging orgy.

The city interior is mostly gray, a dirty clamor of buildings, roads, and sky, occasionally offset by verdant greens that grow rampantly in the subtropical climate and feast on carbon dioxide. The air is gritty with particulates, and one's breath seems to come out in clods. The streets confront pedestrians with a conflict of the senses: beef broths bubbling in sidewalk stalls; a faint chemical whiff, like new-tire smell, emanating from nearby factories; an undertone of sewage. Buckets in underpasses catch leaks just downstairs from bus shelters that advertise the Apple Watch. Car horns are leaned on without relent. Chinese pop music blares from open-doored shops. Busy people in their twenties, with tight jeans and pretty summer dresses, rush from somewhere to somewhere. It is impossible to imagine the sleepy fishing village that this place once was, and it is unlikely anyone bothers to try. Everyone in Shenzhen looks forward.

"Build Your Dreams." There could be no more apt promise for this city in this epoch, but the slogan happens to be under the ownership of BYD Auto, one of the world's largest sellers of electric cars (the largest, if you count hybrids). BYD's auto division has been based in Shenzhen since 2003, when its parent, BYD Company, acquired a failing local manufacturer called Tsinchuan Automobile Company. Its first car, a gasoline sedan called the F3 that retailed for about $10,000, rolled off the production line in 2005.

To get to BYD Auto's headquarters, you have to drive twenty-five miles east from the center of Shenzhen to the industrial suburb of Pingshan, past a dribble of drab manufacturing buildings, bleak apartment complexes with laundry hanging outside the windows, and men selling car seat covers from the side of the highway. En route to my destination, my taxi passed a crane that had dropped a shipping container. Cardboard boxes carrying bottles of motor oil had spilled onto the road.

BYD Auto's campus sits behind an arch of steel pipes and aluminum roofing at its entrance, a faded attempt at industrial grandeur. The sides of the soccer-field-size hexagon that serves as BYD Auto's global headquarters are painted the same sky blue as the shirts that every worker has to wear. My guide for the day, a young woman on the marketing team, was proud to be at BYD, a rare Chinese company that can claim a global presence. Outside of China, it has offices and factories in the United States, Canada, Japan, Korea, India, Mexico, and Europe. Overall, the company brought in about $11 billion in revenue in 2015.

Wang Chuanfu, an engineer and chemist, cofounded BYD in 1995 at the age of twenty-nine with $300,000 in start-up capital he had raised from relatives. The company earned its early fortune by making rechargeable batteries for mobile phones and counted Motorola, Nokia, Sony Ericsson, and Samsung among its clients. After listing on the Hong Kong Stock Exchange and then acquiring Tsinchuan, Wang plotted a course for BYD to become a leading producer of electric cars and solar power systems—a move that would eventually lead to Warren Buffett, through a Berkshire Hathaway subsidiary, buying 10 percent of the company for $230 million. Buffett's investment partner Charlie Munger had told the "Oracle of Omaha" that CEO Wang was like a mix between Thomas Edison and Jack Welch—"something like Edison in solving technical problems, and something like Welch in getting done what he needs to do."

As we walked around the hexagon—a Tetris-era predecessor to Apple's "spaceship" headquarters—my guide told me that she loved living in Shenzhen. The weather's good, and, for a major city, it is accommodating of newcomers, she said. In Beijing and Shanghai, the local governments' anti-speculation policies made it difficult for nonresidents to buy property, but in Shenzhen, it was easy to fulfill their dream of home ownership. The city, after all, is nothing but outsiders—just as eight in ten people are migrants, so are eight in ten home buyers.

BYD's dream is for a zero-emissions world. We completed our circuit of the building—strolling past packed parking lots of BYD cars and dozens of the Denza model, an electric crossover utility vehicle that was a product of the company's joint venture with Daimler—and stepped inside the hexagon. The interior had the feel of a semi-abandoned hospital, with faux marble floors and an almost total lack of natural light.

After a brief tour of BYD's greatest hits in a museum-like show-room for electronics and battery products, my guide led me to a diorama that exhibited the company's vision. Implanted in a desert in the miniature landscape were solar panels and windmills that fed imaginary power to a battery base station on the edge of a green, populated area. A few electric trucks roamed the miniature streets, while a car sat parked in the garage on the ground floor of a luxury house. Electric vehicle charging stations were dotted around like gas stations.

BYD has a mixed approach to electric transport. For private cars, it's not betting on full electrification in the short term. Instead, for the next few years, its SUVs and sedans will mostly be hybrids. The company's belief is that China's charging infrastructure isn't yet ready to support electric cars for most people's living situations. In China's cities—home to 55 percent of the country's population—few people live in standalone houses with their own parking spaces, and most live in high-rises with shared parking lots, or none at all. Plugging in is a problem.

By selling "dual mode" plug-in hybrids, which carry both a battery and a gas tank, BYD still qualifies for government subsidies, which can take as much as $8,000 off the sticker price and, crucially, exempt owners from a license plate lottery that would otherwise complicate their efforts to get on the road. In an attempt to mitigate pollution and control the number of cars that pour onto their roads, local governments in major cities have placed strict limits on who can get a license plate. A new car owner's chances of having their name drawn in the monthly lottery are extremely low. In Beijing in January 2016, for example, the success rate for obtaining a license for a conventional car was 0.15 percent. Those who do get lucky have been known to pay $14,000 at license auctions, a cost that in some cases exceeds the price of the car. To encourage the adoption of new-energy vehicles, however, the governments have waived the license plate lottery system for owners of electric or hybrid vehicles. Unfortunately, what often happens, as Shaun Rein, head of the Shanghai-based China Market Research Group, told me, is that people buy a BYD hybrid so they can get a license plate, but seldom bother to plug it in. Instead, they just rely on gasoline to charge the battery.

BYD is focusing its full-electric efforts on taxis and buses, which, according to the company, account for 20 percent of all vehicle fuel consumption in China. BYD's electric buses are already in operation in downtown Shenzhen and in several US states, including Washington and California. It has electric taxis on the road in Chile, Uruguay, Hong Kong, the UK, and the Netherlands. The company also produces electric forklifts, sanitation trucks, mining trucks, and concrete mixer trucks, which it sells to emissions-conscious businesses and agencies around the world. The California Air Resources Board, for instance, offers grants for companies to make their industrial fleets zero-emission, as part of its California Climate Investments Program. In June 2016, San Bernardino County, one of the state's most polluted air basins, was awarded a grant of $9.1 million to purchase twenty-seven electric trucks made by BYD. The company makes the heavy-duty vehicles for the US market in a factory in Lancaster, California.

While my guide said she liked the company's cars, others I spoke to weren't so sure. A young Beijinger who worked for one of the electric vehicle start-ups that hoped to surpass BYD felt its cars were "cheap but ugly," with outdated functions "not suitable of the needs of young people in the city."

Indeed, BYD's cars are the opposite of sexy. Their exteriors are blocky and their interiors feel plasticky. The company's image is about as staid as the uniform shirts that it forces its employees to wear—and it seems to know. In April 2016, a BYD executive, looking enviously at Tesla, said BYD had made branding its top priority for the next two to three years. "We don't have the ability now to sell tens of thousands of cars before producing a single one," Senior Vice President Stella Li told *Bloomberg*, referring to the Tesla Model 3, which attracted hundreds of thousands of preorders more than a year out from the first deliveries. "The day we can do that will be the day our brand is established." BYD hired a brand consulting firm and made a significant change for a sales event ahead of the 2016 Beijing Auto Show. For his keynote speech, CEO Wang roamed the stage and stressed the company's mission to clean up the air and make roads safer. He was almost like Jia Yueting, or even Elon Musk. The year before, Wang read his speech from behind a lectern.

While BYD attempts to buy market appeal, however, a new generation of Chinese electric car companies are hoping to earn it from the get-go by borrowing some Silicon Valley sizzle.

On a sunny day in May 2016, I walked with Li Xiang up a dusty concrete alley in an industrial district in the northeast of Beijing. There wasn't much in the area except a few car repair shops and the research center for Che He Jia, one of China's most intriguing new auto start-ups. Li, who was also a founding investor in Nio, started the company in 2015 as a thirty-four-year-old and, by the time I met him, had raised $300 million in start-up capital from his founding team, venture investors, and the LEO Group, which specializes in water supply, power station construction, and petrochemical engineering, among other things.

Che He Jia was renting office space in a building in front of one of the car repair shops, so we walked past open garages and crumpled Volkswagens as we headed to the back section. Li walked with a light step in Nike running shoes and blue jeans. As we approached a door in the side of a concrete slab of a building, he waved his hands. "No photos." He opened the door and we stepped inside. The first thing I saw was a Renault Twizy, a buggy-like two-seater with a one-speed transmission and a seventeen-horsepower electric motor. Beside it was an electric moped. I figured they were being used as benchmarks.

I turned to my left and saw three clay models, built to scale, of dinky-looking cars with steep windshields and straight backs. The cars, identical except for minor design variations, were each 3.3 feet wide and 8.2 feet long, with room enough for two people, one seated behind the other. They were painted black and silver, so they looked like mechanical snails from a Daft Punk music video. Their squarish noses added a dash of tough-guy attitude to a dainty physique. This was never intended to be a muscle car, though. Che He Jia calls it a "smart electric vehicle" (SEV) and it's designed purely for city driving, so it has a top speed of forty miles an hour and up to fifty miles of range. It's small enough that four can be (autonomously) parked side by side in a regular parking spot. The company planned to start selling it at the end of 2017.

Che He Jia is one of a rash of new auto start-ups in China, and Li, founder of the publicly listed automotive site Autohome, is one of several Chinese Internet entrepreneurs intent on creating a car company for the twenty-first century. Li and company have been encouraged by the early success of Tesla in the United States, emboldened by government incentives for clean transport, and convinced that the convergence of electric power trains, connectivity, and autonomous-driving technology has created a once-in-a-century opening for newcomers to enter the market.

As well as Nio and Byton, the list of car start-ups funded by Chinese Internet companies also includes Xiaopeng, whose investors include Alibaba, Foxconn, and Russian billionaire Yuri Milner; Sokon, which acquired Martin Eberhard's battery start-up InEVit and named him chief innovation officer at its US subsidiary, SF Motors; WM Motor, started by a former executive at Volvo's Chinese parent company, Geely; Singulato Motors, started by a former Qihoo 360 executive; and a joint effort between Alibaba and the state-owned Shanghai Automotive Industry Corporation (SAIC). Established car companies such as CH-Auto and Changan Automobile are also making electric vehicles. Going by the current birth rate, it's likely that a dozen more electric car companies will have materialized by the time anyone reads these words. In fact, a 2016 study by the China Automotive Technology and Research Center found that the country had more than two hundred manufacturers of new-energy vehicles. Many lagged behind global standards for quality, reliability, and technology. The government has considered imposing strict limits in an effort to improve standards, with one Chinese Communist Party–linked newspaper suggesting that such a move could wipe out 90 percent of the hopeful start-ups.

The central government plays an outsize role in China, where mandates, incentives, and limits are frequently dished out in five-year plans, policy documents, and statements to the press. A company cannot hope to succeed without at least the government's tacit consent—companies are dependent on government-issued operating licenses and other allowances—and it helps to have cordial relationships with the regulators (the Chinese call this *guanxi*). On a macro level, the government controls interest rates, the exchange rate, and the price of energy, among other key levers of the economy. It also has control of major sectors it considers strategic, through state-run oligopolies in banking, energy, and telecommunications, to pick a few.

Electric vehicles fall into a sweet spot of strategic importance (energy) and an industry (auto) already replete with state-owned enterprises, such as BAIC and SAIC. The global ascendance of electric vehicles has also come at a time when the Chinese government is approaching crisis mode over concerns about air quality. The pollution in Beijing is so bad that breathing the air does as much damage to your lungs as smoking two packs of cigarettes a day. Pollution protests have been mounting. At the same time, the government has been attempting to wean the country's economy off its reliance on natural resources—coal, steel, and iron in particular—to instead emphasize innovation. The 2011 Chinese National Patent Development Strategy highlighted seven industries to focus on in the coming decade: biotechnology, high-end equipment manufacturing, broadband infrastructure, high-end semiconductors, energy conservation, alternative energy, and clean-energy vehicles. In 2017, it added artificial intelligence to the list.

New-energy vehicles have been the subject of special attention. In its most recent five-year plan, the government set a goal of having five million new-energy cars on the road by 2020. Accordingly, it promised financial rewards to companies that surpass targets for such things as electric car sales and battery capacity, and it's investing in charging infrastructure while encouraging local governments to offer subsidies to reduce charging fees. It has ordered all government departments to own new-energy vehicles made in China; and it has offered financial incentives to encourage investment in car rentals, battery recycling, and charging infrastructure operations, among other areas. Looking further ahead, government officials have also expressed support for autonomous driving. China is aiming for half of the vehicles on the road to be equipped with advanced safety software by 2020, 20 percent to be highly autonomous by 2025, and 10 percent to be fully self-driving by 2030. It will set a deadline after which automakers must stop selling gasoline cars.

Back in Che He Jia's workshop, Li turned around and walked me over to the SEV's "buck," a polystyrene shell wrapped around a mock-up of the vehicle's interior. I slid into the seat and felt like I was sitting in an arcade-game version of a race car. The steering wheel was a rounded rectangle. There was a touch screen on top of the dashboard, protruding a little into the windshield view. The screen displayed a picture of Taylor Swift, as if an app were playing one of her albums. There were air-conditioning vents, automatic door locks, and a series of controls on the wheel for playing music. I pumped the pedals, imagining I was weaving in and out of Beijing's clogged beltway traffic. It would not be comfortable to be struck by another vehicle—or even an errant watermelon—in this car, but it's unlikely you'd be traveling at high enough speeds to suffer serious injury.

The SEV would be marketed to young consumers in China's major cities—Beijing, Shanghai, Shenzhen, Guangzhou—and was set to retail for about $7,000. As much as it would be a first car for many buyers, it was also like a high-tech upgrade of the electric bikes that Chinese urbanites have been driving for the last couple of decades. Li wanted to supplant the electric bike by providing a low-cost option for people to own a car. "We don't want to challenge Tesla or any other giant automaker," Li wrote on the microblogging site Sina Weibo for his 600,00 followers in October 2015. "We just want to make compact, attractive, and affordable smart cars for everybody."

Li is a multimillionaire high school dropout who grew up under the care of his grandmother in the northern city of Shijiazhuang, in Hebei province. He has always been into technology and an early adopter. As a teenager in the 1990s, he wrote gadget reviews for tech websites and then, as an eighteen-year-old, started his own, PCpop .com. The venture was a success, becoming one of the most well-known electronics review sites in the country and earning millions of dollars in annual revenue. But Li had greater ambitions. At twenty-three years old, he spun off PCpop's auto vertical to create the car information portal Autohome. Autohome started as a simple reviews site like Edmunds.com but evolved into a comprehensive online marketplace that carried independent news and reviews while linking dealers and manufacturers with a trove of consumer data. Autohome quickly became one of China's most trusted sources of automotive information and grew into a highly profitable business. In December 2013, it listed on the New York Stock Exchange for an initial public offering that raised $133 million and valued the company at $3.2 billion, making Li rich. He now lives in a wealthy area called Shunyi, outside Beijing, amid elegant villas and an abundance of Teslas.

Li, in fact, was one of the first nine people in China to own a Tesla Model S. In front of a crowd of reporters and fans numbering in the dozens at a launch event outside Tesla's Beijing office in April 2014, Elon Musk, dressed in a suit and accompanied by his then wife Talulah Riley, handed Li a Model S key. Interviewed by a TV station after the ceremony, a smiling Li mixed praise with skepticism about his new purchase.

"Its external design is that of a million-dollar car. Its driving experience is that of a car costing more than $200,000. But its back seat is that of a $15,000 to $20,000 car." Li later wrote a more damning review of the Model S, praising its smooth driving experience and acceleration but criticizing its leaky sunroof, wipers that made his windshield dirty, "nightmare" rear seats that were too hard, and a substandard interior, which he compared to a Honda Accord's. "There are no cupholders!" he would later joke to me, echoing a major concern of early Tesla owners in the United States. (Tesla now offers cupholder solutions for all its vehicles.)

Like Musk, Li is a hands-on chief executive. At Autohome, he test-drove and reviewed many vehicles himself. His public relations employees at Che He Jia were at pains to stress to me that he has granular knowledge of every aspect of his businesses, and indeed he didn't hesitate to answer each question in careful detail. He is a tycoon in the Silicon Valley mold—geek first, businessman later—and much admired by China's millennials, who see in Li a self-made success willing to turn his back on the country's restrictive education system to pursue his dreams. Unlike many other prominent business figures in China, Li dresses casually and cuts a humble figure, eschewing the self-promoting tactics of, say, Jia Yueting or Zhou Hongyi, the controversial cofounder of Internet security company Qihoo 360.

Li's company prides itself on an Amazon-like understanding of its customers. "We think, 'What does the real Chinese customer need?'" Li said. The electric vehicles the world had seen to date had not been built with the average Chinese consumer in mind, he noted. Tesla's cars were products of California, where there were large highways and long commutes, and people lived in houses with garages that they could charge in. BMW's compact i3, on the other hand, was a product of Europe, where there were condensed urban layouts and short distances between cities, so ninety miles of range was enough to serve most driving needs. In China, by contrast, people were geographically dispersed among hundreds of far-apart but heavily populated cities. China has forty-one cities with more than two million people; more than a dozen cities with a population of more than five million; and five megacities with populations exceeding ten million. Che He Jia's SEV plan made sense for driving within these cities, but the company needed to come up with something else for driving outside them.

Back in the workshop, after I slid out of the buck, Li walked over to a pair of pinboards near the door. He smiled as he showed me sketches of Che He Jia's second vehicle, a long-range SUV planned for release in 2018. It looked pretty badass. It was muscular, with the bulldog nose that was on the SEV, but also speedy-looking, with sharp ridgelines that gave the machine some handsome chops. The side panels came up high to meet squinting windows, and the battery pack sat flat below the floor pan. It was designed for families and luggage, with room to seat five people and an extra storage area in the front trunk. It would be built to withstand a high-speed impact, so people could feel safe driving it long distances on highways.

The SUV won't be purely electric. It will carry a gasoline range extender to allow it to drive 370 miles on a single charge—a concession to China's lack of charging infrastructure. By contrast, the SEV will carry a twenty-two-pound battery pack that can be removed by hand and plugged into a normal power outlet so it can be charged under a desk at the office or at home overnight—a practice familiar to almost every one of the approximately 200 million electric-bike riders in China. The battery pack's lithium-ion cells come from Panasonic, which also supplies Tesla. It takes six hours to charge the SEV's battery fully, Li said, but it can accrue twelve miles of range in half an hour—enough to serve in a pinch.

But maybe people won't have to drive at all. That's the other thing that convinced Li that 2015 was the right time to get into the auto industry. He and his cohorts are convinced that the self-driving car era is imminent. "Autonomy may come even earlier than electric," Li posited. "There's a faster revolution in autonomous driving software than there is in battery technology." Che He Jia was building its cars from the start to be ready for the autonomous era.

I had last been in China during the summer of 2012. While my then girlfriend (now wife) did an internship at a Chinese law firm, I reported on the tech industry and some of the country's most interesting start-ups for *PandoDaily*. After that stay, I wrote an e-book, called *Beta China*, about the emerging innovation culture in China's Internet companies. The premise of *Beta China* was that the country's tech industry was moving out of a phase of copying ideas from the United States and into an era in which it would pioneer products and business models. I highlighted companies such as smartphone maker Xiaomi and apps such as Tencent's WeChat, which has become so pervasive that it is the de facto Internet for many Chinese smartphone owners, as industry-leading innovators that pointed the way to the future.

Four years passed before I returned to the Middle Kingdom. If you have even a passing familiarity with how China's economy has been developing over the last three decades, you might expect me to say here that so much had changed. The cliché about the transformative times the Chinese people have been living through since the country's capitalist revolution is so well worn that it has become as smooth as Yosemite granite.

But still. *So* much had changed.

Viewing Beijing through the windows of taxis—my main form of transportation during my few days in the capital—I saw that many more tall buildings and glitzy shopping malls had been built; that adventurous architects had been emboldened and enabled; and that Western brands like Starbucks and Burberry had increased their visibility, as had local brands like Maan Coffee and Li-Ning. As in Shenzhen, there were advertisements for the Apple Watch anywhere there were shoppers. The air, I might add, seemed a touch more acrid.

But the most significant changes were the things unseen. That was particularly true of the tech industry, which had provided the lighter fluid for an explosion in automotive start-up activity. In 2012, for instance, few Americans paid attention to the WeChat phenomenon. By 2016, however, Tencent had integrated an entire ecosystem of services into WeChat, which was being used on a monthly basis by more than 800 million people. People could not only use the app to talk to friends but also to pay bills, order taxis, buy shoes, and send money to their peers, among many other things. By then, Snapchat founder Evan Spiegel had pointed to Tencent as a "role model," Facebook executive Stan Chudnovsky referred to WeChat as an inspiration for Facebook Messenger, and Kik, a company I worked for part-time, had declared its intention to be the "WeChat of the West." (Tencent later invested in both Kik and Snapchat.)

In the same period, Xiaomi had grown like crazy—in 2015, it sold more than seventy million smartphones—and was briefly (prior to Uber) the world's most valuable start-up before losing some ground thanks to intensified competition in the smartphone market. Meanwhile, in May 2016, Shanghai-made social videos app Musical.ly found its way to the top of Apple's US App Store—the first Chinese start-up to claim the honor—and search company Baidu announced it had developed a self-driving car that had been driving on Beijing's Fifth Ring Road since December 2015.

In October 2014, Shaun Rein, of China Market Research Group, published *The End of Copycat China*, in which he applied an economic lens to the argument I had made in *Beta China*. China had been progressing along an innovation development curve, Rein wrote, from copycat stage to "innovation for China" stage to a global innovation stage. The impetus for the shift toward innovation of global significance, he wrote, "is being driven by Chinese consumers who want the best products and services developed for them directly and by ambitious, well-capitalized companies looking to offset a slowing, more competitive economy by becoming global players."

A day before meeting Li Xiang at Che He Jia's R & D center, I had lunch with one of his new employees, a sales and communications representative who had just finished working for Mercedes-Benz. We ate at an upscale Cantonese restaurant in a shopping mall near some office complexes in the Wangjing district. Between mouthfuls of congee, my host, who was in his early thirties, agreed there had been a cultural shift in recent years. "People are more modest about what they buy now," he said. "They don't want to show off as much." A change in consumption habits had been influenced by a series of factors, notably the central government's crackdown on corruption, slowing economic growth nationally, and the expansion of China's middle class. Now, instead of buying showy luxury goods, many people adopted the postures of Silicon Valley figures such as Musk, Jobs, and Zuckerberg, who embodied power and wealth without having to display it.

Again, Rein covered the subject in *The End of Copycat China*. "Luxury items remained out of the reach of many by the turn of the millennium," he wrote, explaining the earlier era. "Conspicuous consumption and showing off bling became integral to gaining social status." High-glitz brands like Louis Vuitton, Omega, and Montblanc ruled in those days. "The bigger, louder, and more in your face the logo was, the better." Soon, though, as tastes started changing, buyers began to favor craftsmanship, materials, heritage, and durability over conspicuous consumption and labels. "While status remains integral to the [Chinese] culture," Marty Wikstrom, the chair of luxury footwear company Harrys of London, told Rein, "there has been an evident shift in purchasing in order to develop one's own individual and unique style." These were trends that would affect automakers' prospects in the country.

In his office at Che He Jia's R & D center, Li Xiang explained the evolution to me in generational terms, noting its effect on a culture of innovation. Chinese who were born in the 1950s and 1960s, he said, came from austere times and tended to value the best quality at the lowest prices. Those who were born in the 1970s and 1980s grew up with better education and witnessed the economic opening of the country. People from that generation who entered the tech industry looked to Silicon Valley for inspiration and applied the "Copy to China" strategy to serve their home market's needs. (This strategy was so prevalent in China that it got its own abbreviation: C2C.) But those who were born in the 1990s grew up in modern China and would come to develop a markedly different outlook. They had received good educations and been exposed to the Internet from a young age, allowing them to mingle with a global community. Li said they were more independently minded. China's millennials had witnessed the rise of several local Internet giants—such as Tencent, Alibaba, and Baidu—and seen how those companies could compete with, and sometimes even surpass, their American counterparts. Alibaba, for instance, staved off a threat from eBay, and the popularity of its payment service, Alipay, prevented PayPal from getting any meaningful foothold in the country.

It used to be the case that Chinese companies would copy

American websites, Li said. "But now, the US is copying Chinese apps." He probably had WeChat in mind. "Chinese companies and leaders are forced to think of their own patterns of development. They can't follow others anymore."

The development in China's tech industry is borne out geographically. In 2012, ground zero for tech companies in Beijing was the Zhongguancun development in the Haidian district. Home to Lenovo, Microsoft, and Sohu, among many others, Zhongguancun was known as "China's Silicon Valley," a reputation aided by the presence of a preponderance of stores that sold cheap electronics. Zhongguancun is also conveniently close to two top universities (Peking University and Tsinghua University) and the Chinese Academy of Sciences, which have served as important feeders of talent for the tech companies in the area.

By 2016, however, the tech start-up epicenter had shifted. Many of Zhongguancun's electronics stores had disappeared, killed off by the rise of e-commerce. And the coolest Internet companies had moved east to Wangjing, congregating around the Zaha Hadid–designed SOHO shopping complex, made up of three impossible-to-miss curved towers that emerge from the ground like the fins of mechatronic whales.

Zhongguancun was old-school; Wangjing was new-school. At SOHO, you could find hot app companies such as discount shopping giant Meituan and Ele.me, China's largest food delivery start-up. At lunchtime, you would be confronted with a rush of tech workers in their twenties sprawling across SOHO's outdoor plaza in search of food. In the center of the complex was an expansive dancing fountain, with soothing music emanating from hidden speakers. I asked my guide, a young marketing executive at the electric car start-up Singulato Motors, why there was music of mysterious provenance playing so loudly at lunchtime. "Maybe it's to calm down the stressed workers," she offered.

Across the street, Tiger Shen's office inside Singulato Motors' sixteenth-floor headquarters looked down on its own oasis of calm: a lawned garden and a tiled courtyard. Shen, who had been a product executive at the Internet security company Qihoo 360, started Singulato Motors in October 2014 with the Chinese name Zhiche Auto, the translation of which is "intelligent vehicle." I had time to look out his window because he was late for our meeting, having been occupied by talks with government officials. The room was stuffy because the air-conditioning had been off for most of the morning. I was standing just outside his office door, looking at a computer screen that displayed some machine-vision technology, when he hurried in. Shen took off his crumpled beige sport coat, hung it on a coat stand behind his desk, and sat to face me. In an open-plan room next door, eighty-ish workers sat at rows of long tables peering into monitors.

Shen founded Singulato to pursue a market opportunity whose creation he credits to Tesla. "Tesla made the world's first smart car," said Shen, wearing a Tommy Hilfiger blue polo and a mop of hair with ragged strands, like an endearingly nerdy Beatle. He had always been ahead of his class, studying at Shanghai's prestigious Jiao Tong University from age fifteen and graduating by nineteen with degrees in management and automation. In 2000, after moving to Japan to work as an engineer at a software company called Open Net, he started JWord, which solved an Internet search problem for Japanese consumers by replacing user domain names with keywords. In 2005, he sold the company to Yahoo! Japan and became an executive there while taking a stake in Kingsoft Japan, which also specialized in digital security. He helped the company through an initial public offering but later took up an executive role at Qihoo 360, at CEO Zhou Hongyi's request, where he led a team developing connected devices, such as a smart watch for kids and a smart router.

When the Model S arrived in China in April 2014, Shen recognized it as a transformative phenomenon. "I thought it marked a very significant change," Shen said, "like the first iPhone." Tesla applied the software development mind-set of continuous improvement to its cars, and controlled everything digitally, Shen noted. The software became the "brain" of the car and could serve as a platform for other apps in the future. "It was a big change. It's the next generation." He thought there was an opportunity for Chinese companies to do something similar, just as Xiaomi had created a compelling homegrown alternative to Apple. Initially, he wanted to do it within Qihoo 360, but an automotive vertical didn't fit with the company's core mission, which was to focus on security. So he left to start his own company and has since attracted a group of automotive and software veterans to join him.

Since he was a kid growing up on a navy base in Fujian, Shen had wanted to build a car. Now he has one. Singulato Motors crafted a high-concept electric SUV—called the Singulato IS6—that leans heavily into the future. It's a highly stylized beast with a big, beefy nose. The wheels on the concept car had bright orange hubcaps. The company's fans voted for the vehicle's name, which is inspired by "singularity," the science-fiction term that describes the convergence of human and artificial intelligence. The company liked the word so much that it adopted Singulato Motors as its English-language name.

Actually, the vehicle is one of two concepts that Singulato Motors unveiled before asking the public to vote on which one should be put into production. The other SUV, which had Model X–like gull-wing doors, wore narrow rectangular LCD display bands on its nose and rear that would bear programmable digital letters. The show car said HELLO on the front and STOP on the back, to be activated when the driver touched the brakes. Some of the company's employees seemed disappointed that this version didn't prevail in the popular vote.

As with its smart EV contemporaries, Singulato plans to make its cars autonomous and does not expect to make much money selling the hardware. Instead, it hopes to generate revenue from other services and products, like insurance, maintenance, charging, and smart parking (Che He Jia plans to do something similar). The IS6, for instance, could theoretically communicate with a smart charging station so that it could park and then charge itself. The charging service provider would automatically connect to the car to manage and personalize the process, while handling the billing accordingly.

If this model proved prescient, it would be bad news for traditional automakers, for which software has never been a core competency. And indeed, Shen's prognostication for the old guard is not optimistic. If the world does get to a future of self-driving electric cars that can talk to one another and the infrastructure, I asked him, where would gasoline cars be in twenty years? "There will basically be none," he answered evenly. He leaned forward and brought his hands together on the table so that they formed a triangle. Tesla's Model S already had a driving range of three hundred miles per charge, he said, and the battery technology would only improve. He predicted a four- or fivefold increase in range for electric cars over the next twenty years, coupled with a steep decrease in costs. At that point, there'll be no place for gasoline cars in the market.

"It's certain to happen," he concluded.

<hr/>

While serious upheaval in the auto industry may well be imminent, it would be foolish to ignore the possibility that Singulato Motors, and every one of its young contemporaries, could fail. After all, three out of four venture-backed start-ups in the United States fail to return their investors' money. In China, the odds aren't necessarily any better. It's harder to find data on start-up failure rates there, but, given that seven new companies start every minute in China, it's safe to assume that there are a few misfires. Perhaps Singulato is naive about its ability to capitalize on a software–based business model.

Perhaps Che He Jia, in launching its SEV within two years of the company's founding, is being too aggressive with its timeline. Perhaps Nio, in attempting to launch its cars globally almost simultaneously, is being overly ambitious.

None of these companies has proven definitively that it can merge the disparate cultures of the automotive and technology industries into a single entity. None has yet developed a car that can fully drive itself, and no one knows exactly how China's regulators will respond to the prospect of autonomous transport.

The charging infrastructure challenge also remains an open question. As of 2016, grid operators in China were running far behind targets for charging-station installations and hadn't found a way to make them profitable. The Chinese economy, hampered by slower growth, has also been badly saddled with debt. Its total debt in 2015 was 250 percent of its gross domestic product, according to a government economist, a result of the Chinese Communist Party's growth stimulation programs to shore up the slowing economy. How long can it hold? Any serious recession, or worse, could wipe out the venture capital–dependent auto start-ups. And a more complicated trade relationship with the United States under President Trump only makes the future murkier.

For Tesla, which bears a large part of the credit for encouraging this surge of start-up activity, the issue is less about imminent failure and more about how pronounced its success could be.

China is vital to Tesla's long-term prospects—and indeed, the future of electric cars. If China doesn't go electric, then Tesla's goal of accelerating the world's transition to sustainable transport will be undermined. Meanwhile, to state the obvious, the sales potential in the world's largest auto market, which still has plenty of room to grow, is enormous, and electric vehicle sales will be helped by the government's mix of incentives and subsidies. Tesla has said that the country could account for a third of its car sales in the long term. It could well be more. But Chinese consumers did not initially show great willingness to adopt electric cars. Even though electric car sales in China increased 223 percent in 2015, they still represented only 1.4 percent of the market—and that included hybrids.

For a while, it looked like Tesla's impact in China would be underwhelming. China, with unique challenges from governmental and cultural quarters, is a difficult place for American tech companies to do business. Many have come and failed, including Google, Yahoo!, eBay, Groupon, and Uber, to name a few of the most prominent. Such companies have been brought low through a combination of hubris, lack of familiarity with the business climate, and misjudging local tastes. In some cases—Google's, especially—they haven't worked well with the Chinese government (the *guanxi* weren't adequately lubricated).

There were some early signs that Tesla was making some of the same mistakes. For example, it initially overlooked wealthy Chinese consumers' unique tastes. In particular, as Li Xiang suggested, the back seats in the first Model S delivered to China were bench-like and uncomfortable, which was a problem in China, where wealthy citizens often prefer to be driven by chauffeurs. In early 2015, Tesla addressed the problem by offering an "executive" rear seat upgrade that effectively converted the bench into two armchairs. Meanwhile, some of the features that were standard in Tesla's cars in other markets were missing in China. For instance, for the first few months, Chinese Model S drivers were unable to access the navigation system on the car's maps. The feature wasn't added until early 2015.

Other unanticipated problems also emerged, including an issue with scalpers, who attempted to buy the Model S in bulk so they could on-sell the cars at a markup. A sketchy businessman tried to hold Tesla ransom by registering the company's name before it entered

China. Cars got stuck in customs. Making matters worse, unlike other automakers, Tesla also lacked a local manufacturing partner, which meant its cars didn't qualify for some government subsidies.

The company's China efforts, according to *The New York Times*, were off to a "lurching start." Musk would later concede that sales had been "unexpectedly weak."

In response, Tesla replaced a country manager who had been in the job for less than a year, and laid off as much as 30 percent of its local workforce. "I think Tesla took for granted that they were just going to succeed in China," Ricardo Reyes, a former vice president of communications and marketing for Tesla, told *Fortune* in June 2017.

But all was not lost. In fact, on reflection, it looked like Tesla was following a similar path to the one taken by Apple, which also got off to a rocky start in China. Apple launched the iPhone in China in 2009 but was struck by a series of criticisms early on, not the least of which were about working conditions in the Shenzhen factory where its devices were made. The factory owner, Foxconn, came under fire for several worker suicides. Meanwhile, Apple's scalper problem was on another level. The scalpers were so bold as to buy iPhones in bulk and sell them right outside Apple stores. Apple was also slow to make the latest iPhones available in the country, with Chinese consumers typically getting them months after they went on sale in the United States. And for the first four years that the iPhone was on sale, it wasn't even compatible with China Mobile, which accounted for 700 million customers.

Since then, Apple has fixed those problems, and China is now arguably its most important market, having registered more iPhone sales than in the United States. In October 2015, Apple CEO Tim Cook said China will be "Apple's top market in the world." Apple's commitment to the country shows in its products. In particular, the large-format plus-size iPhones (starting with the iPhone 6 Plus) were perfect for China, where mobile phones serve as showpieces. The flashy Apple Watch is an equally powerful status item.

There is little reason to suggest that Tesla can't capitalize on the same dynamics. Tesla has made moves to bolster sales in China, including a deal with China Unicom to install charging stations at retail outlets, and it has committed to spending hundreds of millions of dollars on service centers and charging installations around the country. At the launch of the Model X, Musk made a big deal of the car's "hospital-grade" cabin air quality, a desirable feature in smoggy Chinese cities. Tesla has also opened more stores in the country, and it has signed a deal with the city of Shanghai to build and operate a factory in the city—a move that will help it not only get its cars to market in China more quickly but perhaps also get on the list for more government subsidies.

Following Apple's lead, Tesla is adapting to the local conditions.

In June 2016, the company's vice president for Asia Pacific, Robin Ren, announced that Tesla had completed its hundredth Supercharger station in the country, making it possible to drive two thousand miles from the northern city of Harbin to the southern city of Shenzhen relying on Superchargers alone for energy. Tesla had also partnered with local utility companies to install chargers in homes and in parking lots, and it had installed 1,400 fast-charging posts at shopping malls, office buildings, and hotels.

By the end of 2016, Tesla had tripled its sales in the country, despite a year and a half of bad press, well-publicized charging challenges, zero paid advertising, and a presence in only seven Chinese cities. The Model 3, meanwhile, could find a market in China's still-growing and tech-savvy middle class, which may not be able to afford a luxury car but could be willing to stretch to premium mass market. In the wake of the Model 3 announcement in March 2016, China was the second-largest market for reservations.

There is hope.

On my last night in Beijing, I met two friends for dinner at a roast duck restaurant in the Parkview Green mall, an enormous glass-walled pyramid in Beijing's central business district. As well as being notable for its environmentally friendly credentials, the mall was the site of Tesla's first store in China. It had opened in November 2013.

After picking through Peking duck at our patio table on the warm spring night, I decided to go see the Tesla store. One of my dinner companions came with me and we walked along a darkened hallway past shuttered luxury retailers—Van Cleef & Arpels, Alfie's Beijing, a high-end sushi restaurant—before coming upon the store, which was lit up like a beacon. It stretched most of the length of the mall's southern wall.

"You can't go in there; it's closed," a security guard told us as we approached.

"It's okay, we just want to take a look from the outside," my friend said.

It was three days after the fiftieth anniversary of the beginning of the Cultural Revolution, a social upheaval that had inflicted trauma on the rich and conspicuously educated. The *People's Daily*, a government mouthpiece, marked the occasion by calling the Cultural Revolution "a mistake . . . that cannot and will not be allowed to repeat itself."

China had learned its lesson from that decade of tumult, the paper continued, and was determined to avoid any social unrest that would disrupt progress. "The Chinese people have never been so close to realizing the goal of the great rejuvenation of the Chinese nation."

From the dim corridor, we peered into the lighted room and saw two Model S's parked on an epoxy floor, surrounded by an array of branded T-shirts and white logos. We lingered for a few moments, just the two of us and a perplexed security guard, looking in a window to an empty store. This was new China: savvy, capitalist, and hopeful. Tesla's presence was still that of a leader from a foreign land, but its biggest challenges could soon come from within the country.

For the time being, my friend and I looked in at the preeminent symbols for a future of clean transportation and wondered what story would be told of them in the decades to come. The cars—vehicles for a new revolution—looked proud but lonely.

10

A SUPERTANKER'S
THREE-POINT TURN

"I'm not really a fan of disruption."

Simon Sproule, a spry British communications professional, worked at Tesla for eight months in 2014 before leaving a week from the launch of the Model S P85D. He had decided to join his friend and former colleague Andy Palmer at Aston Martin. Palmer gave Sproule the responsibility of running marketing communications and installed him in a spacious glass-walled office at Aston Martin's headquarters in Gaydon, England, ninety miles northwest of London on the M40. The well-put-together Brit had left the brave new world of Silicon Valley for the decidedly old-school confines of one of the world's most iconic car brands.

At Tesla's headquarters, a visitor must drive into the parking lot off Deer Creek Road, just outside Palo Alto, hand their keys to a valet, then sign in on an iPad in the reception area. For my visit to Aston Martin, I drove down a private road called Kingsway, informed security guards at a checkpoint that I was there to see Sproule, and was then directed to the VIP entrance, where I pushed a buzzer by a large locked gate. Walking past a private lot of million-dollar cars, I entered a brightly lit lobby with a high ceiling under which had been parked a series of Aston Martins—the Zagato, the Rapide, the Vulcan—just waiting for 007 to come in and select one for his next impossible escape across the continent.

Sproule, who was recovering from a cold and sported Pierce Brosnan–level stubble, greeted me and escorted me to his office. A suit jacket was draped over his chair, and he wore a sweater over a crisp-collared shirt.

Aston Martin's sales had been in a steep decline when Palmer was appointed CEO in September 2014, and as he assumed the leadership, it continued to bleed money. In October 2015, the company announced that it had lost almost £72 million the previous year. It would cut 15 percent of its staff. Palmer said the company wouldn't get to profitability until 2017. This was not a particularly unusual state of affairs. Aston Martin had had multiple owners and several near-death experiences since it was founded by Lionel Martin and Robert Bamford in 1913. It went into, and out of, bankruptcy in the 1970s and enjoyed one of its best periods of stability under the ownership of Ford, from 1991 to 2006. By 2016, it was owned by an international consortium of investors, including the Italy-based Investindustrial and two Kuwait-based investment funds.

Palmer and Sproule were attempting to turn around Aston's fortunes, with the ultimate goal being an IPO or a sale to a larger automaker. Thanks in part to Mr. Bond, the company had retained its brand cachet, but it had done little else to promote itself. In a very British manner, it had erred on the side of discretion. "That's fine," Sproule said as he sat opposite me at his glass-topped conference table, "but we were very elegantly going out of business." Sproule said his new boss knew that Aston Martin had to be "in the conversation." That was something that Musk was extremely good at. "Everyone knows Tesla now," Sproule said, somewhat in disbelief that a small auto start-up could attract so much attention. "It's electric cars. How the fuck did that happen? How did that happen?"

Sproule's job was to keep Aston Martin in the headlines—a task at which he was proving adept. On his desk that day was a copy of the *Financial Times*, which was reporting that there was a "heavily over-subscribed" waiting list for a hypercar that Aston was working on with the Red Bull racing team: "the most expensive road car in British history." Aston had also been in the news for its plans to build a £200 million factory in Wales for its upcoming crossover utility vehicle, the DBX, a departure from the sports car line on which it had built its reputation. But the news that had brought me to this pretty spot in the English countryside was that Aston had teamed up with Jia Yue-ting to build an electric car. In February 2016, the company announced that it had signed an agreement with LeEco to develop and produce the RapidE, an electric version of Aston's Rapide, a high-performance se-dan. They planned to get it to market by 2018.

The RapidE would be the vehicle that allowed Aston to learn how to make an electric car, Sproule told me. He conceded that it would be inherently compromised because it wasn't a "ground-up electric car," but Aston had to start somewhere. The electric era was coming and Aston needed to be part of it. "The brand is handcrafted British excellence," Sproule said. "It's beauty. It's power. None of that is mu-tually exclusive with electric cars."

At that moment, Andy Palmer walked in and shook my hand. Sandy-haired and heavy-chinned, Palmer slid his thick-stock business card across the table. At the top of the card was the Aston Martin logo, embossed in charcoal type with the company name set against a set of stylized wings—an homage to the scarab beetle, revered in ancient Egypt as the embodiment of a god who pushed the sun across the sky.

Palmer, in his early fifties, had come to Aston Martin after twenty-three years at Nissan, where he had been chief planning officer and head of the luxury brand Infiniti. He served under Carlos Ghosn, the respected but feared CEO and chairman of the Renault-Nissan Alliance, but ultimately found he had risen as far as he could. The Brit had started his automotive career as a sixteen-year-old apprentice in England's West Midlands and, since he turned twenty, dreamed of being CEO of a car company. The role at Aston Martin, just five miles from where he went to high school, seemed made for him. Indeed, the move was likely more than pure coincidence—Palmer had reportedly tried to convince Ghosn to buy a stake in Aston Martin in 2012 and 2013. While at Nissan, Palmer had promised to the press that the company would be "the absolute, number-one leader in zero emissions," largely on account of the lead it had built with the Nissan Leaf. He now credited Nissan with making the modern electric car a commercial reality, and Mitsubishi with being the forerunner of the electric era with its i-MiEV. "Tesla was a slow third," he added, apparently disregarding the role of the Roadster.

Palmer, who slumped comfortably in his chair, wearing a blue suit and no tie, agreed that he was preparing Aston for an electric transition. It was essential for automakers to master electric vehicles if they wanted to be relevant in 2025, he said, even if that meant losing money at the start. But he wasn't convinced that the industry's newcomers— such as Faraday Future, Nio, and Byton—were up to snuff. "They all have this innate belief that the auto industry is stupid, it's slow, it's a mammoth," he said. They believed that they could come in, change the business model, create connected electric cars, change the sales process, and change the way cars were manufactured. "That's also bollocks," Palmer concluded, steadily holding my gaze. "Complete bollocks."

"Why?" I asked.

"Well, you can't ignore a hundred and twenty-five years of history. The auto industry didn't not learn anything." When you put together a car, most of it is mechanical, and most of it is done by people who have spent a career working out how to do a suspension system, or a door lock, or a steering wheel, Palmer said. "And you can't ignore that. So you've got to buy it. And you either buy it through a consultancy, or you buy it through recruitment, or you buy it through collaboration." This spelled trouble for the newcomers. "The majority of those start-ups will fail because they'll be too slow to recognize that they can't just trash the auto industry, that there's something relevant that they need."

So why was Jia Yueting different?

"He's not," Palmer responded flatly. "I mean, he's a very modest guy, but his dream is the same. These guys dream of changing the world, they dream of taking their Internet model and bringing it into the auto business. In that respect, he's not that different, and his probability of success is about the same."

So why agree to work with him?

"Because I think the industry needs some of these guys to be successful. And China is different from the United States insofar as it doesn't have a successful auto industry, and therefore successful auto engineers." Jia doesn't know what he doesn't know, Palmer said, and that meant Aston had an opportunity to help.

And what of Tesla? Did he think it would fail?

Of all the start-ups, Palmer conceded, Tesla was the one most likely to make it, simply because it had come so far already. But its funding was tenuous and it didn't have enough money in the bank to think beyond the Model 3. "That's the story of Aston Martin. Why aren't we successful? It's because we never make enough money to pay for the next car. That's exactly what Elon is facing now." The challenge would only intensify as Tesla developed the Model 3 in a segment where margins are razor thin and the costs of globalizing the product would be immense. "It's going to get harder, not easier."

And then there was Tesla's most serious challenge. The company's success had been "pretty phenomenal," Palmer conceded, but—and here he slipped into a coarser tone that the British would recognize as working class—"it still can't make any doors that work, can it?" At the time, the Model X was experiencing reliability issues, with many early owners reporting problems with the falcon-wing doors, among other flaws. "Still can't make trim that doesn't fall to pieces," Palmer continued. "Still can't put a piece of rubber around a door."

Of all the criticisms Tesla has faced in its history, the one that has carried the most weight is that it has not mastered manufacturing. In 2016, it delivered about seventy-six thousand vehicles, missing its original target of eighty thousand to ninety thousand for the year. As well as the falcon-wing door troubles in that year's batch, there had been a recall for an issue with the third row of seats, which Tesla found could fold over in a crash. The Model S had seen its share of problems, too. It had early defects with the auto-retracting door handles, frequently failing drive units, and various other niggles throughout the vehicle. The problems were such that, in October 2015, *Consumer Reports* withdrew its "recommended" designation for the car after its survey of 1,400 owners found "an array of detailed and complicated maladies." According to its ensuing report, "The main problem areas involved the drivetrain, power equipment, charging equipment, giant iPad-like center console, and body and sunroof squeaks, rattles, and leaks." In other words: pretty much everything. The next year, Tesla's reliability improved to an "average" rating, earning back its *Consumer Reports* recommendation.

Tesla applies a philosophy of "continuous improvement" to its manufacturing process, meaning it keeps on developing its cars even after sales have commenced. There have been signs that, partly as a result of this process, the company has been getting better at manufacturing. "Reflecting our philosophy of continuous improvement, we have not relaxed our pursuit of making the world's most reliable cars," Tesla said in a letter to shareholders in February 2016, noting that the cost of repair claims had halved in a year. Still, it is not a good sign to have struggled so much with production quality even at low volumes. A company that produces only tens of thousands of cars a year must be considered niche when compared to automakers that produce millions. Toyota, General Motors, and Volkswagen, for instance, each produced about ten million cars in 2015. Palmer, then, had reason to be skeptical about Tesla's ability to fix its quality issues as it attempted to scale to a production rate of half a million cars a year by 2018—and into the millions beyond that.

Considered alongside other auto industry veterans, however, Palmer came across as a veritable Tesla fanboy. Many of his peers have expressed considerably more skepticism. "There is nothing Tesla [can] do that we cannot also do," Fiat Chrysler CEO Sergio Marchionne (who died in 2018) said in June 2016. Two years earlier, he had asked customers not to buy the Fiat 500e electric car, because the company lost $14,000 on the sale of each one. Fiat would sell the minimum number of electric cars needed to meet government mandates and "not one more," he said. In April 2016, Marchionne continued that theme in an interview on the sidelines of his company's annual meeting, this time responding to the price of the Model 3. If Musk could show him that the car would be profitable at the $35,000 price tag, Marchionne said, "I will copy the formula, add the Italian design flair, and get it to the market within twelve months." The German automakers have been even more dismissive. In November 2015, Edzard Reuter, the former CEO of Daimler, called Tesla a "joke" and Musk a "pretender," suggesting in an interview with a German newspaper that Tesla didn't stand up to serious comparison with "the great car companies of Germany." Daimler, BMW, and Volkswagen were slow to accept that Tesla could one day challenge their market dominance. "German carmakers have been in denial that electric vehicles can create an emotional appeal to customers," Arndt Ellinghorst, an automotive analyst at Evercore ISI, told the *Los Angeles Times* in April 2016. "Many still believe that Tesla is a sideshow catering to a niche product to some tree-hugging Californians and eccentric US hedge fund managers." GM wasn't quite so blasé. In 2013, then CEO Dan Akerson established a team within the company to study Tesla, based on the belief that it could be a big disrupter. GM's Chevrolet Volt, a hybrid sedan that could drive about forty miles in full electric mode, had won *Motor Trend*'s 2011 Car of the Year, but GM was looking further into the future. At the 2015 Detroit auto show, it unveiled a concept of the Chevy Bolt, a two-hundred-mile electric car that would retail for $30,000 (after a $7,500 rebate from the US government). It was seen as a direct response to Tesla and new CEO Mary Barra's biggest risk since she took over in 2014. *Wired* magazine celebrated the Bolt's impending arrival with a February 2016 cover story about how GM had beaten Tesla "in the race to build a true electric car for the masses." The cover showed Barra in a leather jacket, arms crossed, posing beside the Bolt, and the tagline: DETROIT STRIKES BACK!

In July 2016, I spoke to Pam Fletcher, GM's executive chief engineer for electrified vehicles, about the company's electric-mobility strategy. Fletcher is a GM veteran who spent her childhood in small-town Sarahsville, Ohio, at the racetrack with her father. After graduating with a master's degree in engineering from Wayne State University, she went on to work for GM, Ford, and then a company that developed NASCAR engines. She rejoined GM in 2005 and became part of the engineering team for the Chevy Volt, distinguishing herself as the chief engineer for the propulsion system, and then head of department for the Bolt. Fletcher said she couldn't imagine a more exciting time to be working in the industry. The trifecta of electrification, autonomous driving, and car-sharing programs had convinced her that "we are literally on the cusp of the biggest transformation since the invention of the automobile."

The Bolt is just one of GM's two dozen vehicle models in the United States, and Fletcher referred to it as a component of the company's expanding "mobility play." In 2016, GM acquired the self-driving car start-up Cruise for a billion dollars and invested about half that amount in the ride-sharing company Lyft, giving the auto giant a firm footing in the trifecta of areas that Fletcher said was transforming the industry. "The Bolt EV is the tipping point that kicks all that off," she said.

Because we spoke long in advance of the Model 3's public availability, Fletcher wouldn't be drawn out on the question of whether the car was a Model 3 competitor ("You're asking about something that I don't know what it is"), preferring to focus on bigger-picture matters. GM has to cater to a diverse array of customer concerns around the world. "The variety of needs is so great that you're going to see a number of solutions," she said. "The propulsion that goes with them will evolve over time." By the same token, she suggested that any wholesale transition to electric mobility is likely to be lumpy if you look at the issue from a global perspective, where different regions have different problems to solve. Consequently, she thought the internal combustion engine still "has a while" left in its life span. On the other hand, certain parts of the world are ripe for change. "If you start breaking it down more finely, and you start looking at specific megacities or specific city centers, and look at regions that have common issues, you may see a transformation happen in pockets at a much more rapid pace."

The Bolt is a practical car with good handling and a range of 238 miles per charge as assessed by the US Environmental Protection Agency. Some observers, such as the *New York Times* tech columnist Farhad Manjoo, have suggested that by beating Tesla to market with a long-range mass-market electric car, GM bested the Silicon Valley company at its own game ("How did GM create Tesla's dream car first?" Manjoo asked in a piece heralding the Bolt's arrival). But it's unlikely Musk and company are too worried.

Tesla has long said that it wants to accelerate the world's transition to sustainable transport, and a good part of that means encouraging other automakers to do more with electric cars, and faster. "We hope the big car companies do copy Tesla," Musk said in 2014. "I don't know why they're taking so long." If anything, the advent of the Bolt is evidence of Tesla's success in advancing the agenda for electric mobility.

When asked for her thoughts about Tesla, Fletcher applauded the company's efforts: "Adding great entries and awareness into that marketplace is exactly what's required."

Compared to the mass-market electric vehicles that have come before it, the Bolt is a master achievement. It boasts more than twice the range of the 2016 Nissan Leaf (although an upcoming version of the Leaf looked set to drive two hundred miles per charge) and beats the seventy-six-mile-per-charge Ford Focus Electric on most measures. It boasts about three times the range of a BMW i3 and is $6,000 cheaper.

I test-drove a Bolt in December 2017. It was a pleasant ride, like being in an extremely peppy Toyota Yaris, but with more headroom. The clearest evidence I was in an electric car came when I put it into "Low" mode and felt the full force of the regenerative braking. When I stomped on the accelerator, the Bolt gave me a gentle jolt and took me to sixty miles per hour before I remembered that I was supposed to be counting the seconds. It was a compact car, but there was plenty of room for passengers. The flat floor gave the vehicle a sense of openness, and the trunk was deep and spacious enough to comfortably accommodate two golf bags.

While Fletcher would not compare the Bolt to the Model 3, it's fair to ask how it measures up to Tesla's long-range entrant in the same price category. The Model 3, with a fastback design and storage space under the hood, looks more like a BMW 3 Series than a Bolt, which in style is closer to economical compact cars like the Yaris or the

Honda Fit. Other features, like the touch screen UI or the access to charging stations, don't bear comparison.

Some critics, including former GM chairman Bob Lutz, have wondered if the Bolt is as much about lowering average fleet emissions to comply with government fuel economy mandates as it is a legitimate attempt to commit to electric mobility. Theoretically, at least, the zero-emissions car allows the company to get away with selling its more profitable but more polluting SUVs, such as the Chevy Tahoe, and pickups, such as the Chevy Silverado, a good while longer. It also seems significant that GM officially calls the car the Bolt EV, as if the qualifier is crucial to its distinction in the market. The name suggests that the company sees the Bolt as serving an "EV market." Tesla, by contrast, has long said it's not competing against other electric cars—it's competing against gasoline cars.

A look at production numbers for the Bolt might hold the answer. In the first year of sales, Chevy sold just over twenty thousand Bolts, which is a good number but still meager for a company that sold 9.8 million cars in 2015. It's also a small number compared to Tesla's plans for annual production of the Model 3, which it expects will soon be in the hundreds of thousands.

Confirming GM's relative lack of enthusiasm for the Bolt, Chevy's marketing director said at the Los Angeles Auto Show in November 2016 that brands like the Bolt are "important to our image, but they kind of live on the fringes when it comes to volume." That same month, *The Detroit News* reported that GM expected to lose $8,000 to $9,000 on every Bolt it sells.

There are other questions, too, not least among them being whether GM dealers will fully embrace the opportunity to sell the Bolt. When I test-drove the car, the manager at the dealership told me the Bolts were selling so quickly that most of them didn't last a day on the lot. But we were just outside San Francisco, ground zero for electric cars.

Chevy sent most of its Bolts to Northern California, the dealer said, and a few to Oregon and Arizona. The Bay Area was the only place in the country where the Silverado pickup wasn't Chevy's top-selling vehicle. Instead, the hybrid Volt held that title. So, when the "brown suits" came down from Detroit, the dealer said, they came with smiles, thanking the Californians for their good work selling the low-emission vehicles because it allowed them to keep selling lots of pickup trucks and SUVs.

One also has to wonder if GM will build a serious charging network. When I asked about charging options for the Bolt, I was pointed to paid charging posts in Target parking lots and charge points at Chevrolet and Cadillac dealerships across the country. By contrast, Tesla had about eight thousand Supercharger points and fifteen thousand fast-charging points worldwide by the end of 2017. Fletcher did not outline any specific plan for charging infrastructure but said the company will work with consumers to see what they want. She noted that GM's electric car customers mostly charged at home or work anyway.

Still, it would be unfair to completely downplay the significance of the Chevy Bolt. Here is a quality electric car at a competitive price from one of the largest automakers in the world. Because of the Bolt, more people will hear about electric cars, more people will drive them, and more people will have an alternative to burning fossil fuels to power their primary mode of transportation. Even the most ardent Tesla fans and the EV1's most morose mourners must salute that.

——————

Most automakers seemed relatively comfortable with their place in the world as they moved into the mid-2010s. Electric cars had come and gone in the past, and those that survived—the Leaf, the i-MiEV, the Model S—claimed only a tiny portion of the market. The year 2015 was bringing record car sales as the industry continued to recover from the doldrums of the recession, and gasoline prices were low. It seemed that nothing could convince them that a wholesale strategic rethink was in order.

Well, nothing short of a crisis.

Just before 4:00 P.M. on September 23, 2015, three gray-haired men in spectacles and suits stood with grim faces in front of a Volkswagen signboard at the company's headquarters in Wolfsburg, Germany. Berthold Huber, in his midsixties, bald, and with a severe expression, was the first to speak. "The Volkswagen supervisory board learned of the manipulation of emissions from diesel motors with great dismay," said Huber, the board's acting chairman. He paused to cough into his hand. To his left was fellow board member Stephan Weil, the governor of Lower Saxony state, which had a 20 percent stake in the company. To Huber's right stood Martin Winterkorn, Volkswagen's sixty-eight-year-old CEO, who had been in the position since 2007. As the biggest crisis in the automaker's history unfolded, Winterkorn had refused to resign. "We are aware not only of the economic damage that it has caused, but also above all else of the loss of confidence among many Volkswagen customers worldwide," Huber continued. He was reading from a piece of paper as cameras clicked around him. "We are in agreement that the situation needs to be clarified and that all offenses are punished." Winterkorn, who in 2015 was the highest-paid executive in Europe, looked at the floor. "At the same time, we are resolved to embark with determination on a credible new beginning." As Huber announced that Winterkorn was stepping down, the departing CEO contorted his lips and momentarily opened his mouth, as if the air hurt to breathe.

Five days earlier, the US Environmental Protection Agency had ordered VW to recall 482,000 diesel cars after finding that the automaker had used software to cheat emissions tests. VW later admitted that it had fitted eleven million cars with programs—called "defeat devices"—that would self-activate to limit nitrous oxide emissions during the EPA's tests, which took place on rolling roads in controlled facilities. When independent researchers conducted tests on open roads, they found that VW's cars emitted up to forty times more toxic fumes than permitted.*

In the first week after the revelations, VW—which also owns Audi, Bentley, Bugatti, Lamborghini, Porsche, SEAT, and Škoda—lost a third of its value. It was the start of a long and painful process that would ultimately cost the company many billions of dollars. In 2016 and 2017, it agreed to pay more than $23 billion to settle lawsuits in the United States and Canada, $10 billion of which would be paid to affected customers. As part of the settlements, it set up a $2 billion fund dedicated to the advancement and adoption of zero-emissions vehicles. In January 2017, the FBI arrested a VW executive in the United States in connection with its ongoing investigations and charged him with conspiracy to defraud the country. In 2018, Winterkorn was charged with fraud by the US Justice Department.

Since the scandal emerged, it has widened to cast suspicion on numerous other automakers, including GM, Mercedes-Benz, Mazda, Honda, Fiat Chrysler, Renault, and Opel, all of which have been, or continue to be, the subjects of investigations, legal complaints, or recalls. Alarm bells rung the loudest, though, at Mitsubishi Motors, which in April 2016 admitted it had been improperly testing the fuel economy of its cars since 1991. The next month, Mitsubishi's president, Tetsuro Aikawa, and executive vice president Ryugo Nakao, announced they would step down from their positions. In late 2016, the suspicion extended to Volkswagen's Audi vehicles after testing by the

---

* The discrepancy was discovered by the US-based International Council on Clean Transportation, which commissioned a real-world driving study by a research team from West Virginia University. The study was completed in 2013 and corroborated by the EPA.

California Air Resources Board was said to have found software in a recent model that masked the true extent of carbon dioxide emissions.

Despite the wide-reaching ramifications, there is a chance that the Volkswagen-instituted crisis will one day be known for a more important contribution to history than condemning a formerly great company to temporary ignominy. The side effects for electric cars could be more significant than the $2 billion fund that VW has been compelled to establish. After all, a good crisis should never go to waste.

On June 16, 2016, the media were back at VW's Wolfsburg headquarters for another press conference, this time headed by newly appointed CEO Matthias Müller, who had previously been CEO of Porsche. Standing behind a table with another Volkswagen-branded signboard behind him, Müller, white-haired and lean, told reporters that the company had initiated the biggest transformation in its history. VW would reshape itself to become one of the world's leading providers of sustainable transport. "The term *evolution* would be too weak for what we're facing," Müller declared.

Three weeks earlier, following GM's investment in Lyft and on the same day that Toyota announced a strategic partnership with Uber, VW had announced that it would invest $300 million in the ride-booking app Gett. Within days of that news, the German media were reporting rumors that VW planned to spend up to $11 billion on an advanced battery factory that would rival Tesla's Gigafactory. (In March 2018, the company announced that it would spend $25 billion to secure batteries for electric-vehicle production at sixteen of its factories.)

In Wolfsburg, Müller vowed to foster a culture of innovation, which included plans to hire another thousand software engineers to work on autonomous driving features, connectivity, and battery technologies. Over the next decade, VW would launch more than thirty electric vehicle models, and Müller predicted that by 2025 a quarter of the company's sales would be electric. By then, it expected to be making them at a rate of between two million and three million a year.

Müller, who would ultimately be pushed out of his job by VW's board, spoke of taking advantage of the "huge opportunities" presented by the "revolutionary transformation."

Much had changed in the nine months since Winterkorn resigned. GM had shown off its first working prototypes of the Chevy Bolt, Tesla had unveiled the Model 3, and the world's largest seller of automobiles had gone from looking back on the ruins of a fossil-fuel-instituted disgrace to looking forward to an electric-oriented future in which software and sustainability would be at the heart of the company. Historians of the future may well see here an inflection point that determined the fate of the electric car. But those same historians may wonder what took VW and its cohorts so long. Big Auto shouldn't have needed a crisis to realize that the resurgence of the electric car posed an existential threat to its business.

A satirical review published on Tesla Club Sweden's website in April 2015 highlighted the forgotten absurdity of the gasoline car. "The car's gasoline engine coughed to life and started to operate," wrote the reviewer, Tibor Blomhäll, of his test drive.

"One could hear the engine's sound, and the car's whole body vibrated as if something was broken, but the seller assured us that everything was as it should [be]. The car actually has an electric motor and a microscopically small battery, but they are only used to start the petrol engine—the electric motor does not drive the wheels. The petrol engine then uses a tank full of gasoline, a fossil liquid, to propel the car by exploding small drops of it. It is apparently the small explosions that you hear and feel when the engine is running."

Blomhäll went on to note that the engine consisted of "literally hundreds of moving parts that must have tolerance of hundredths of a millimeter to function," that "gasoline engines apparently cannot be driven as smoothly as electric motors," and that the front of the car was "completely cluttered with hoses, fittings, fluid reservoirs, and amid all a huge shaking cast iron block which apparently constituted the motor's frame." To top it off, the car had a chimney that spewed noxious gases into the air everywhere it went.

Blomhäll's review highlighted how strange it is to assume that the internal combustion engine—invented around the same time as the first telephone message was sent, as Edison came up with the phonograph, and as the world's first typewriter punched out its first letters— should prevail as motor transport's primary technology. The auto industry, however, has habituated itself to overlooking the internal combustion engine's vulnerability. For the most part, it has instead projected that vulnerability onto Tesla, pointing to its issues with fires, or finances, or subsidies, or sales. "I assume that [Tesla] will fail because the company has huge losses every year, and there's not enough market demand," automotive consultant Professor Fritz Indra told *Handelsblatt* in 2015.

The critics have a point. Electric cars still need time to win over the public. Even though charging networks are growing by the day, range anxiety remains an issue for the general consumer, as does the cost, even at $30,000 for a midsize car. Not everyone believes that doing away with the roar of an internal combustion engine is a good thing, and when gas prices are cheap, fewer people may be attracted by an electric car's promise of affordable fuel. The financial equation also remains tenuous for all players involved. The big automakers still haven't figured out how to make money from electric cars, and Tesla, if it makes one big bet too many, could yet face financial ruin.

However, those same critics have paid scant attention to the shortcomings of gasoline cars. Even if gas were a dollar a gallon, it would still be significantly more expensive to fill the Mercedes-Benz S-Class's tank than it would be to charge a Model S for equivalent mileage—and

you'd still have to go to a gas station to get the job done.* Gasoline
engines would always be relatively noisy, which some people love but
others would trade for the silence of an electric car in a heartbeat.
Conventional cars would still need regular oil changes and frequent
maintenance to make sure their thousands of parts all work in harmony. (As automakers add technology to their vehicles to reduce
emissions, they are becoming only more complex.) And they would
still be wholly reliant on an economically volatile resource that, when
burned, contributes to the warming of the Earth's atmosphere.

Gasoline cars, meanwhile, will never enjoy numerous other advantages inherent in electric vehicles, like the extra space freed up by the
lack of an engine and driveshaft, or instant torque, or the smooth
driving experience enjoyed as a result of never having to search
through gears. The average efficiency of an internal combustion engine in converting fuel into a car's forward energy ranges from about
14 to 30 percent. For the electric car, it's about 90 percent.

But the real difficulty for anyone arguing the case for gasoline cars
is found in the economics. We are fast approaching a time when gasoline cars will no longer be able to compete with electric cars on price.

To date, the number one factor holding Tesla back from offering
cheaper cars has been the energy cost per unit of its lithium-ion battery packs, which is why it started by selling only high-end vehicles in
which the cost of the battery could be absorbed by the premium price
point. Tesla has never revealed exactly how much of its cars' costs can
be attributed to the battery pack, but in 2013, chief technology officer
JB Straubel told the *MIT Technology Review* that it accounts for less
than a quarter of the cost of each vehicle—which for the eighty-five

---

*A Mercedes-Benz S-Class has a twenty-one-gallon tank and goes 420 miles on a full
tank. The eighty-five kilowatt-hour Model S can use eighty kilowatt-hours of its battery, with 95 percent efficiency from the charger, and goes about three hundred miles
per charge. The average cost of electricity in the United States, according to the Department of Energy, was $0.12 per kilowatt-hour in 2015. It would thus cost $15 to fuel
an S-Class for three hundred miles and $10 for a Model S.

kilowatt-hour Model S, at that time, would have put the battery pack somewhere in the $18,000 to $25,000 range (assuming Straubel was factoring feature-rich versions of the car into his calculations). That would have put the cost per kilowatt-hour of the battery pack at anywhere between $210 and $300. The good news for Tesla is that lithium-ion battery prices, as was inevitable, have been on the way down for years now. Despite some uncertainty because of fluctuating costs of battery materials such as lithium, cobalt, and nickel, the downward trend will likely be accelerated by Tesla's Gigafactory, at which the company has claimed it will be able to reduce pack costs by a third simply by consolidating production processes under one roof and benefiting from economies of scale. Prices go down as processes get more efficient and reproducibility goes up.

Here's where things get interesting. Analysts have found that electric cars will reach cost parity with equivalent gasoline cars when battery cell prices hit $100 per kilowatt-hour. (Note that the component costs of a Tesla battery pack add about 20 percent to the overall cost of producing the individual cells.) In April 2016, Tesla's vice president of investor relations told analysts that the base version of the Model 3 would be offered with a battery pack of lower than sixty kilowatt-hours, and that the cost of a pack was already below $190 per kilowatt-hour. That price was well ahead of analyst predictions, and it has ensured that Tesla can sell an attractive $35,000 electric car with more than two hundred miles of range. But battery costs still have a long way to go. Soon, $190 per kilowatt-hour will be seen as expensive.

For the last two decades, lithium-ion batteries have been facing downward price pressure from the proliferation of laptops and smartphones. It took just fifteen years for the cost of laptop battery cells to drop from $2,000 to $250 per kilowatt-hour, according to Deutsche Bank—an improvement rate of 14 percent a year. From 2010 to 2016, that rate increased to 16 percent a year, according to calculations done by Tony Seba, a clean-energy investor and author of the 2014 book

Clean Disruption of Energy and Transportation. It's unlikely that rate will slow down anytime soon. Indeed, in 2017, the price of a lithium-ion battery pack was 24 percent lower than the year before, according to a Bloomberg New Energy Finance survey.

Price pressure is now coming not only from makers of smartphones and laptops but also from makers of electric cars and energy storage systems. "It's an order of magnitude more demand that we'll see from vehicles and the grid than what people used in their cell phones and laptop computers," Straubel said in 2015. "So that is going to drive down the prices much faster than I think most people expect." He said Tesla would be disappointed if its battery costs weren't in the "ballpark" of $100 per kilowatt-hour by the end of the decade.

Even if Straubel's estimates prove overly optimistic, battery prices will fall to $100 per kilowatt-hour by 2023 just by following the 16 percent per year cost improvement that the world saw between 2010 and 2016. And that's probably a conservative estimate. GM has predicted that its lithium-ion cell costs will hit $100 per kilowatt-hour by 2021.

Keep in mind that these cost reductions require no breakthrough in battery technology, and they don't take into account improvements likely to arise from increased competition, consolidation, scale, and innovation as automakers and utilities push further into the market. The effect will be dramatic. Even with conservative calculations,* a Model 3–quality car could go on the market in 2023 for $25,000. "Starting in 2025, it will make no financial sense to purchase a new gasoline car in any market," former Cisco executive Seba wrote in *Clean Disruption.* As a clean-energy investor, Seba has an interest in coming to such conclusions—but his calculations are straightforward

---

* "Conservative" in this case assumes that the battery pack accounts for 20 percent of the cost of a Tesla, that a Model 3 could use a fifty kilowatt-hour battery pack, and that it would take seven years for Tesla to go from $190 per kilowatt-hour to $100 per kilowatt-hour.

and based on raw data. "Even assuming that my predictions are off by five years or that it takes an extra five years to build out the manufacturing infrastructure and transition to the new EV world order, gasoline cars will be the 21st century equivalent of the horse carriages by 2030."

Unfortunately for traditional automakers, it's not just "cheaper and better" that they should worry about. It's everything about the way they work.

If traditional automakers are accustomed to five-year product cycles—meaning each model is substantially redesigned every five years—how can they keep pace with the rate of change set by a company that is committed to "continuous improvement"? The top-of-the-line Model S came out in June 2012 with a zero-to-sixty-miles-per-hour time of 4.2 seconds, an eighty-five kilowatt-hour battery pack, and no access to Superchargers. By 2017, it was available in dual-motor mode with all-wheel drive, a hundred kilowatt-hour battery pack, zero-to-sixty acceleration in 2.28 seconds, access to more than five thousand Superchargers worldwide, and a technology package that allowed the car to drive itself on the highway.

On battery technology, other automakers have a long way to catch up to Tesla, which has its own battery production facility and a development head start of at least four years. Then, what do other automakers know about delivering software updates to their customers over the air? GM has said it will bring over-the-air updates to its general fleet "before 2020." But what advantages have the incumbents ceded to Tesla while it has been collecting and learning from fleet data since the Model S hit the roads in 2012?

No electric all-wheel-drive car has been put into production by any company other than Tesla. No car company has a charging network that comes close to being as extensive as the one Tesla has been working on since 2012. In the United States, no automaker has been able to sell directly to consumers or establish its own Apple-like retail stores. None has the extent of electric-vehicle service knowledge that Tesla has.

And what about the employees? If you're an engineer who sees electric cars as the future, do you work for a large automaker that has a sideline in electrics, or do you go to a company that's wholly committed to the cause? If you're a software developer, do you want to work for a legacy company based in Detroit or a tech company in Silicon Valley? Can a legacy automaker set in its ways replicate the culture of innovation that Tesla and other start-ups live and breathe? Which car company can match Tesla's tech-brand cachet? Which one has a leader as notorious as Elon Musk? Is there any other single person in the auto industry who can command an instant audience of many millions from a Twitter account?

Maybe the question of whether Tesla can challenge Big Auto is the wrong one. Perhaps the better question is: Can Big Auto survive?

In one of my conversations with Martin Leach, the NextEV cofounder and auto industry veteran told me that the traditional automakers were inherently resistant to reinvention. "They can't really change," Leach said. "They're structurally disabled from making the sort of changes that I would like to see." If you're a newcomer, you don't have that incumbent problem, he added, so you can look at the business differently.

"You often hear the supertanker analogy," Leach said. "It's difficult to turn a supertanker, and I think it's absolutely the case with car companies." Even if the CEO of a big car company decided that he wanted it to go 100 percent electric tomorrow, he would struggle. Internal political battles would prevent the boss from taking the necessary action. "I can't imagine how he would go about that," Leach said. "It would be so difficult."

I asked Leach for an example of how internal politics made radical change difficult. He offered an anecdote from when he took over as president of Ford Europe. After assuming the role, he established a team to set up a management information system that would allow him to monitor critical details about the business. "The first thing that happened is, when they went to everybody to get the information, people basically refused to give the information. Most people would say, 'Yeah, but we don't really want Martin to see that information raw until we've had a chance to look at it and if necessary sanitize it' or whatever—which is code for manipulation." It took Leach five months to put the system in place. After he eventually parted ways with Ford, the project manager told him the system was scrapped within a month of his departure.

Leach also saw a challenge for incumbents with legacy infrastructure. The big automakers have factories around the world working on internal combustion engine cars. BMW has thirty manufacturing sites in fourteen countries. They've got dealership agreements, established supply chains, and long-baked logistics arrangements and practices. There are commercial contracts to honor, relationships to look after. "They just can't pick it all up and rotate it," Leach said. Imagine all the jobs that would be lost. Imagine the equipment suddenly made useless.

Leach suggested that time is not quite up for the traditional automakers, but the clock is ticking. "If we go out in the twenty- to thirty-year time frame, I don't think the auto industry is going to look anything like it looks today."

------~~~~~~~~------

Musk was looking tired as he sat in one of those white faux-leather armchairs that seem standard for conferences. "A lot of people think I'm a fan of disruption, but I'm not really a fan of disruption." He let out a short, juddery laugh-hahhuh. "I'm just a fan of, like, things being better." He had just been asked by Ted Craver, chairman of the utility industry's Edison Electric Institute, to comment on the notion that Tesla might be a disruptive force.

"Probably for a lot of people in our industry," Craver had said while fiddling with his pen, "they hear this word *disruption*; *transformative business plans*; *disruptive technologies*; and I think at the core of that is a sense of the word *displacement*." It was June 7, 2015, and the men were onstage at the Hyatt Regency New Orleans with JB Straubel, at the institute's annual convention. "I think the audience would be interested in hearing your views on this 'disruption' and how these companies are going to behave with each other."

Musk looked ceilingward before delivering his answer. "I think that if there's a need for something to be disrupted and it's important to the future of the world, then sure, we should disrupt it. But I don't think that we should just disrupt things, unless that disruption is going to result in something fundamentally better for society."

The business world had come to fear the word *disruption* because of *The Innovator's Dilemma*, a 1997 book by Harvard Business School professor Clayton Christensen that introduced the concept of "disruptive innovation." *The Innovator's Dilemma*, lauded by *The Economist* as one of the most important books about business ever written, lays out how market-leading companies can miss out on new waves of innovation by focusing on the needs of their existing customers to maximize profits while overlooking new and cheaper technologies or business models that customers didn't realize they wanted. Disruption theory explained why Kodak went from industry ruler to also-ran, why Amazon surpassed Barnes & Noble, and why Netflix obliterated Blockbuster.

The tech start-up world from which Musk hails embraces disruption as one of its organizing principles, encouraged in part by the influential blog *TechCrunch*, which named its flagship conference, TechCrunch Disrupt, for the concept. Silicon Valley's budding capitalists have long been encouraged to use their software prowess and processes to disrupt existing industries, and hence we have Facebook, which disrupted the news media industry, Airbnb, which disrupted hotels, and crowdfunding, which disrupted traditional investing. When Ted Craver asked Musk to share his thoughts on disruption with an audience of old-school electricity providers, you could see why the chairman might nervously fiddle with his pen. Could Tesla, with its emerging energy-storage business, disrupt the utilities?

It might have come as some comfort to those at the conference that Musk is no fan of disruption. Indeed, he and Straubel were probably there to convince utilities to work with Tesla on energy storage projects that could benefit both parties. But the industry's fear that it might have been on the wrong side of history would not have dissipated completely. The same was true for at least one auto industry leader.

The man who, until May 2017, was CEO of the Ford Motor Company is one person who does appear to be a fan of disruption. Mark Fields, a Harvard business grad and Clayton Christensen follower, was fifty-three when he was appointed to succeed outgoing CEO Alan Mulally. Fields had been the company's chief operating officer since November 2012, and his ascent through the ranks had included leadership roles at Ford's luxury autos group and Mazda when Ford owned the company in the early 2000s. In 2006, he was given the task of shaking up Ford's corporate culture, which employees had described as "toxic," "cliquish," and "hierarchical." He attempted to foster "a sense of crisis, but not panic" among employees, according to *The Wall Street Journal*. In meetings, he would often utter the imperative "Change or die."

In April 2016, Fields gave an interview to the tech news site *The Verge*, which introduced the CEO with the warning that the combination of autonomous vehicles, ride-sharing, and Tesla posed a larger threat to the auto industry than it had ever confronted before. Fields, student of Christensen, was ready. He told the publication that Ford's approach was "to first disrupt ourselves."

"You know, when I first joined the company, a long time ago, we were a manufacturing company," Fields said. "As we go forward, I want us to be known as a manufacturing, a technology, and an information company. Because as our vehicles become a part of the Internet of Things, and as consumers choose to share their data with us, we want to be able to use that data to help make their lives better."

Fields noted that Ford had announced FordPass, a mobility solution that he hoped would "do for the auto industry what iTunes did for the music industry." The service would let Ford owners reserve parking, share cars with others, and pay bills using a Ford payment system (FordPay). "Let's say you're going through a drive-through, and instead of taking your wallet out, your vehicle will know that it's you, and you'll be able to pick up your McDonald's and will get billed through FordPay." Unfortunately for Fields, his vision did not translate into value for shareholders. Over Fields's three-year tenure, Ford's shares dropped 40 percent. In May 2017, the company replaced Fields as CEO with Jim Hackett, who had been responsible for Ford's autonomous driving efforts.

To realize its forward-looking vision and become a leader in automotive technology, Ford would need the services of the world's best software developers, which would mean competing not only against other automakers but also against Silicon Valley's hottest companies. In the new era of automotive, software is king. With that shift comes an opening for software-focused companies like Tesla. "In many cases, large car companies or truck companies are not focused on software, they're not focused on sensors or batteries," Straubel said in 2016. "And this gives an opportunity for innovation for new companies and new entrants to play on a bit more of a level playing field than there ever was in the past."

Tesla, however, is not a disrupter by Clayton Christensen's original definition of the theory. In a December 2015 article published by the *Harvard Business Review*, Christensen and his coauthors, Michael Raynor and Rory McDonald, said Tesla didn't fit the disruptive innovation model because its foothold was in the high-end segment of the auto market. Far from representing a marginal technology, Tesla's entry to the market had elicited significant attention and investment from established competitors. "If disruption theory is correct," Christensen concluded, "Tesla's future holds either acquisition by a much larger incumbent or a years-long and hard-fought battle for market significance."

Tesla may not fit Christensen's classic definition of disruption, but that doesn't mean the auto industry shouldn't be concerned. After all, the theory has shown its limitations in the past. In 2007, five months after the iPhone was announced and eleven days before it hit stores, Christensen predicted the smartphone wouldn't succeed. "They've launched an innovation that the existing players in the industry are heavily motivated to beat," he said in an interview, adding that history speaks loudly on that point and "the probability of success is going to be limited."

Tesla doesn't seem to care about the semantics. At a conference in November 2015, Straubel said the company doesn't even think about disruption. "We don't sit around thinking of 'How can we disrupt something else today?'" he said. "That's never happened inside of Tesla. I've never been in a meeting where disruption was the focus of the meeting." Instead, Tesla just thinks about how to best make use of available technology to make an amazing product. "It's really that simple. We just look at the customer and the technology, we put them together, we try and invent a new product, innovate new features around that."

One can get too tangled up in theory. Christensen's "disruption" might have its own specific meaning, but the dictionary definition of *disrupt* offers a broader understanding: "to cause (something) to be unable to continue in the normal way." With that, we can safely put aside our copies of *The Innovator's Dilemma*. Its rules need not apply here. We may be about to see the iPhone effect applied to a trillion-dollar industry.

March 31, 2016. Elon Musk is standing on a black-floored stage in a black suit over a black T-shirt, in black shoes. His hair is a perfect whoosh of brown. He got here at 8:50 P.M., only twenty minutes late. Behind Musk, whose jacket collar has been propped vertical, is a wide-screen projection that spans the length of the stage, which itself stretches across the Tesla Design Studio's concrete floor. By the time he gets to talking about the Model 3's features, Musk has already given the audience a brief history of the company, explaining why it started with high-end vehicles so that it could fund the development of a mass-market car such as the one about to be revealed. Musk speaks faster and louder than he has at previous appearances. He stumbles over his words less.

Safety typically comes first in Musk's presentations. "The Model 3 will not just be five-star on average," he announces. "It will be five-star in every category." While he's saying this, a 3-D rendering of the car's chassis spins on the screen behind him. As it turns, the animation builds out its body, so that by the end of the safety spiel, the car's skeleton is on full display. The audience, an invited list of Tesla owners and fans that numbers in the high hundreds, is slow to start clapping, but then comes a wave of applause and whistles.

"Even the base Model 3 will do zero to sixty miles an hour or zero to a hundred kilometers an hour in less than six seconds," Musk declares, raising his voice for those last four words. "At Tesla we don't make slow cars." The crowd laughs. He promises that, of course, there will be versions that go much faster. "All right!" shouts a man in the crowd.

"The range will be at least an EPA rating of two hundred and fifteen miles," says Musk, eliciting another loud cheer from the crowd. "I want to emphasize, these are minimum numbers. We hope to exceed them." "Go, Elon!" someone cries.

Tesla started accepting reservations for the Model 3 at its stores that morning. Prospective owners had to part ways with $1,000 to place the fully refundable order, even though they knew essentially nothing other than the car's price. People arrived at the stores early, in some cases the night before, forming lines reminiscent of the hordes of excited Apple fans that gather to buy iPhones on the day they become available. Model 3 deliveries would not start for at least another eighteen months.

In Short Hills, New Jersey, more than two hundred people were in line before the store opened its doors. One of those customers, a twenty-seven-year-old athletic trainer, had been there since 2:30 A.M., sleeping in the mall parking lot. At 7:50 A.M., there were 340 people in line in Austin, Texas, and hundreds outside the showroom in Bellevue, Washington. At 10:00 A.M., there were 300 people in line at Tesla's Palo Alto store. There were similar scenes in Denver, Raleigh, Seattle, Zurich, and in pouring rain in Montreal.

In Boston, an Audi A4 owner had camped outside the store in a tent. "I'm very much a traditional car enthusiast," he told *Jalopnik*. "But to me, Tesla is the future." A musician waiting outside Tesla's store in Red Hook, Brooklyn, said he had been waiting all his life for something like this to happen, even though he had never liked cars. A software engineer eager to place his order recoiled when asked if he was interested in the Chevy Bolt. "A Chevy Bolt? Oh God, no. No. The Chevy Bolt's like a Compaq; this is a MacBook. That's the difference."

Back at the launch event, Musk continues his sales pitch. "All Model 3s will come standard with Autopilot hardware." No one would have to pay extra for Autopilot's safety features, such as automatic emergency braking and blind-spot detection. "The Model 3 also fits five adults comfortably," he says. The car's outline, its interior stripped, is displayed on-screen. People lift their smartphones to take photos.

"The rear roof area is one continuous pane of glass. The reason that that's great is because it gives you amazing headroom and a feeling of openness." Musk is sounding like Steve Jobs. "It has by far the best roominess of any car in this size." The car has trunks in the front and rear. "It has more cargo capacity than any gasoline car of the same external dimensions."

Supercharging? That comes standard. The crowd applauds ecstatically. On the wall behind Musk, specks of light glow on the surface of planet Earth as seen from space. By the end of 2017, Tesla will have more than seven thousand Superchargers and fifteen thousand destination chargers in its network, Musk promises. The cars' onboard chargers will automatically adapt to any voltage in any country.

Deliveries will start at the end of 2017, Musk says, before adding, with a sheepish grin: "I do feel fairly confident that it will be next year." The audience, knowing Musk's punctuality issues, laughs heartily.

The price? Thirty-five thousand dollars, of course. Even the base version of the car will be better than anything else on the market in its segment, Musk promises. "You will not be able to buy a better car for thirty-five thousand dollars, or even close—even if you get no options." He is not smiling now.

"So do you want to see the car?"

The crowd, loose with liquor, roars in unison: "*YEAHHHH!!!*"

"Well," Musk starts, "we don't have it for you tonight"—he leans on the last syllable and puts his hands up in front of him as if in apology. There are groans from the crowd, then laughter. "I'm just kidding, of course. It's April Fools' somewhere! All right, let's bring them out!"

The lights go down. Two and a half seconds of silence pass, and then comes epic orchestral music fit for a *Star Wars* fight scene. A video starts playing on a screen the size of a wall. It shows the curves of a red car in close-up. A Tesla badge is on the hood. The door handles, flush with the panels, look like capital *L*'s lying on their side. Ensuing shots show the chrome of a hubcap, headlights like the eyebrows of a Vulcan, and an indented nose.

The video hands off to reality. Three Model 3s drive out: red, silver, and matte gray. Lights sparkle off the windshields. The cameras, on booms, peer inside. There's a horizontal touch screen in the minimalist interior, and white leather seats. Cupholders! The crowd is allowed a minute to gorge on the visual feast.

Three weeks after the unveiling, Musk would tell a conference in Norway that the Model 3 was a "real wake-up call" for the auto industry. "No one seemed to care when the Roadster was launched."

Big Auto has indeed now woken up. Volvo has announced plans to sell only electric or hybrid cars from 2019 onward. BMW will produce an electric version of its bestselling 3 Series . . . sometime around 2020. Its CEO, Harald Krüger, said in October 2016 that the company "will systematically electrify all brands and model series." In April 2016, Daimler shareholders raised concerns that the company had no competitor to the Tesla Model 3. The company followed by announcing the "EQ" electric vehicle sub-brand at the 2016 Paris Motor Show, an effort that will include the rollout of a charging network. Daimler is spending $10.8 billion to make at least ten electric models available by 2022. VW has said it will invest $50 billion in five years on electrification. Nissan has released a next-generation Leaf that competes with the Model 3 for range. Ford said it will invest $4.5 billion by 2020 to develop thirteen electric vehicles, including a new SUV with three hundred miles of range, and a hybrid version of the F-150 pickup truck, the bestselling vehicle in the United States. Even Toyota, which once placed all its alternative-energy faith in the hybrid Prius and hydrogen fuel cells, changed its mind and decided to start making long-range electric cars, including as part of a joint venture with Mazda. GM has declared that it will ultimately go all electric and has promised twenty new electric models due by 2023.

The supertanker is attempting a turn.

At the Tesla Design Studio, Musk walks back out after the parade of the new Model 3s. "So what do you think?!" he shouts. He's pumped up now. "You like the car?"

Arms are up, grins are out, and the crowd is going bonkers. But there's one more thing.

"This is kind of crazy," says Musk, scratching his head, "but I just learned—was just told—that the total number of orders for the Model 3 in the past twenty-four hours has now passed a hundred and fifteen thousand!"

By the end of the week, a stampede of orders would account for $14 billion in implied future sales. Tesla would declare the event "the biggest consumer product launch ever."

PART THREE

THE OPEN ROAD

11

ELECTRIC AVENUE

"The world does not lack for automotive companies."

Boyd Danks is seventy-six years old and works as a barman at the USA Tavern, a windowless room within a gas station adjacent to the Tahoe-Reno Industrial Center in Nevada's Storey County. The bar has slot machines built into the counter and a pool table in the corner, around which are gathered three burly men in T-shirts, lazily knocking back pitchers of Modelo. It's 1:00 P.M. on a Wednesday and pushing 90 degrees in the high desert outside, but I'm with Danks in the darkness, nursing a Corona and listening to his thoughts about the Model 3, two weeks after Musk delivered the first cars to customers.

Danks is not particularly impressed.

"Yeah," he says, his voice slightly raspy with age, "the standard version is thirty-five thousand dollars, but if you want it in any color other than black, that's another thousand dollars, so that's thirty-six thousand dollars. And if you want the version with three hundred and ten miles of range, that's another nine thousand, so you're looking at forty-five thousand dollars—and even then, it can't get me to Vegas." Las Vegas is 413 miles away from Danks's hometown of thirty years, Fernley, which itself is a twenty-minute drive east from the USA Tavern. "And still no one's told me how long it takes to charge one of those things." He's happy with his Nissan pickup, never gonna bother with an electric car.

Danks, who is Q-tip bald and used to work on oil rigs in Montana and Wyoming, probably knows more about the Model 3 than the average small-town septuagenarian, but that has a lot to do with his bar's location. It sits at the top of Electric Avenue, the service road off Interstate 80 that leads to what will be the world's largest lithium-ion battery factory. The USA Tavern is the nearest watering hole to the Tesla Gigafactory.

Danks might not be a fan of the Model 3, but he can see some of the benefits that the factory, still three years from completion, has already brought to the area. The state government has almost completed work on a new four-lane road that connects I-80 with US Highway 50—providing a shortcut from the industrial center to the town of Silver Springs—a project that Danks says the authorities have been talking about for twenty-five years and has been helped along by the Gigafactory's presence. The factory has also brought the area thousands of new jobs, many of which required skills not possessed by locals, so people have come from all over the country—Florida, Missouri, Alaska—to fill the need. Some of them drink at the tavern. On the other hand, rents have gone up. One of Danks's friends recently vacated an apartment in Fernley for which he paid $850 a month. The new tenants pay $1,100.

More jobs, increased property values, and high-tech innovation— these are the treats that Governor Brian Sandoval promised Nevadans when he brokered a deal with Tesla to bring the Gigafactory, which would be built in partnership with Panasonic at a cost of $5 billion, to the state. When Tesla announced a short list of five states for potential

Gigafactory locations in early 2014, there was intense competition to win the electric car company's favor. Rick Perry, then the governor of Texas, drove into California's state capital in a Model S and said he wanted to see a MADE IN TEXAS bumper sticker on the car. Arizona promised to put forward a bill to allow Tesla to sell its cars direct to consumers in the state. New Mexico considered a special legislative session to pass Tesla-specific incentives. California state senator Ted Gaines personally delivered a glitter-covered "golden shovel" to Tesla's headquarters.

But after months of wrangling, it was Sandoval who ultimately prevailed, thanks to an assortment of incentives that could be worth about $1.4 billion over two decades. Tesla got 980 acres of land for free, a twenty-year exemption for sales taxes on equipment and construction materials, ten-year exemptions for property and payroll taxes, $8 million in electricity discounts, and $195 million in transferable tax credits. In return, Tesla promised 6,500 jobs and a $35 million contribution to Nevada's education system. The Nevada site also held appeal because it's only a 235-mile drive to Tesla's Fremont vehicle factory and a twenty-minute drive from downtown Reno, with easy access to rail and highways. Corporations in the state also enjoy zero income tax.

Musk first spoke publicly in August 2013 about the need for a "truly gargantuan factory of mind-boggling size" for battery production. Speaking to CNBC's Phil LeBeau, Musk said there wasn't enough lithium-ion cell supply to satisfy Tesla's long-term needs, so the company had to do something about it. The "giga" part of the name indicated a scale factor of billions, which was exactly what would be required to enable Tesla to produce half a million electric cars a year. Initially, it planned to build a ten-million-square-foot factory that would produce thirty-five gigawatt-hours of energy capacity a year by 2020, but Tesla later revised that figure to fifty gigawatt-hours by 2018. The factory would not only serve Tesla's supply need, the company explained in a blog post, but also reduce the cost of lithium-ion cells by eliminating waste, consolidating production processes under one roof, and benefiting from economies of scale. By the time the factory was running at full speed, it alone would produce more batteries than all of the lithium-ion battery manufacturers combined in 2013. When, on September 4, 2014, Musk and Sandoval announced the deal on the steps of the state capitol in Carson City, it was greeted as a solemn occasion. It was, Sandoval said, a "significant and historic day" that would "change Nevada forever." Musk applauded Nevada as a "real get-things-done state."

Three years on, the promised changes were starting to take effect—and not just in the immediate surroundings of the USA Tavern. I had retired to the tavern after a lunch at the Subway in a small strip mall next door that was full of customers wearing Tesla T-shirts and security badges. In the parking lot were eight Model S's and Model X's, sitting outside an unmarked Tesla service center.

But signs of the Gigafactory's economic impact were obvious even in Reno, a city known chiefly for its garish, and fading, casinos. I stayed at a place that boasted of being the city's first nonsmoking hotel. It had a climbing gym on the second floor and an upscale bistro downstairs. When I checked in, a sign by the elevators advertised an orientation for new Tesla employees. The next day, I shared an elevator with a man carrying a hard hat and wearing a Tesla security badge on his belt. At breakfast, a gaggle of Panasonic employees in yellow safety vests lined up at a buffet. A server told me that more than two hundred of the hotel's 350-odd rooms were set aside for Gigafactory workers.

A few streets over, a food and beverage enclave that was part of an urban revitalization project peddled goat cheese and rosemary ice cream, farm-to-table kimchi tacos, and kava kava from coconut shells in a Fiji-themed bar. While the staff at these places held the Tesla people in some suspicion—the Gigafactory employees appeared to them almost cult-like in their devotion to the company's mission—they acknowledged that the arrival of the Muskites helped make their fledgling businesses viable. Meanwhile, even the Silver Legacy casino—"Experience luxury in downtown Reno"—had caught Tesla fever. Parked atop a row of slot machines on the casino's ground floor was a Model S 70 that was proffered as a grand prize in a "spin the wheel" game. GO BIG, read the license plate.

My last night in Reno coincided with the first day of Hot August Nights, an annual festival that bills itself as the largest nostalgic-car show in the world. Even as an automotive luddite, I appreciated the alluring curves and throaty growls of the Thunderbirds, Corvettes, and Mustangs that rolled under the Reno Arch on the city's main street. As rain fell in the twilight, it struck me, in an obvious, high school poetry kind of way, that the juxtaposition of these titans of the gasoline era with the Teslas parked outside the Gigafactory twenty-three miles away told a pat story about the past and future of the automobile—that today's cars are tomorrow's relics, that the thousands who came as tourists for the display of the internal combustion engine's fullest glory were outnumbered by the thousands who had arrived to build an electric future. But it was also hard to picture a scenario in which a hobbyist's love for an electric car would have quite the same animal effect as a man's love for his painstakingly restored ride. A large part of the satisfaction to be gleaned from classic-car ownership comes from the perfect purr of the internal combustion engine, tuned just so, tweaked precisely to an individual's taste. An electric car might be able to match a gasoline car's curves, but it can never match its soul.

I was surprised to find myself thinking these things, especially given that I have no love for gas guzzlers. But perhaps I felt a sense of imminent loss—that these nostalgic-car meet-ups might be the only way my kids and grandchildren will have any inkling of the automotive era in which I grew up. Even as I knew I had seen the future, I felt stuck in the past.

Then, a dark green Model T puttered by, a reminder of how this all started. How much of the world that car changed. The roads, the garages, the urban development, the planned communities, the population mobility, the advent of megacities, the burned oil, the mass production, the new American economy, the eight-hour workday. The Model T had a soul and a momentous story, the story of modern civilization.

That story couldn't have been told without Henry Ford's invention, in 1913, of the moving assembly line, an innovation that dramatically sped up and reduced the cost of manufacturing the Model T. The moving assembly line brought the price of the Model T down from $850 in 1908 to $360 in 1916, helping to push the vehicles out of the realm of the elite and into the garages of the middle class. A century later, Tesla's Gigafactory, essential to its mass-market hopes, stood to have a similar catalytic effect on the spread of electric vehicles. Soon, a great electric car would be affordable for the average consumer. How much of the world could the Model 3 change? Perhaps it, too, will be rolling under the Reno Arch a hundred years hence.

But that factory at the end of Electric Avenue is about more than just cars. It is the anchor weight for a revolution that has significance well beyond the replacement of gasoline. The Gigafactory is the inverse of the oil rigs that Boyd Danks used to work on—not a facility that drills deep into the earth to extract fossil fuels but a building on top of it to prevent the need for them. It is a powerhouse for a new energy economy.

————◦◦◦◦◦◦◦◦————

It's 9:20 P.M. on April 30, 2015, at Tesla's Hawthorne design studio, and Musk has walked out to techno music almost an hour after the advertised start time. He's wearing a gray suit jacket and black shirt. His first words to the audience are "All right." Directing the crowd's attention to the screen behind him, he shows photos of power plants spewing dark smoke from their chimneys. The concentration of carbon dioxide in the Earth's atmosphere is a threat to humanity, he says. If we don't stop burning fossil fuels, we will see carbon dioxide concentration at "levels we don't even see in the fossil record." He shows the Keeling Curve, which plots the frightening escalation. A man in the crowd shouts: "Save us, Elon!" Musk hesitates and laughs. "I think we collectively should do something about this," he offers, "and not try to win the Darwin Award." (The Darwin Awards "salute those who improve the species by accidentally removing themselves from it.") Then Musk gets on with his message. He wants to talk about a missing piece.

There's not enough wind or solar power generation today to replace fossil fuels completely. A major barrier has been that renewables are inconstant and therefore unreliable. The wind doesn't blow all the time, and the sun doesn't shine at night. Musk says that there is a "fairly obvious solution" to these problems. We need better batteries.

Tesla already makes battery packs for its electric cars, and those packs function as energy storage units, holding power pumped into them from the grid. Tesla, as Musk explains, decided to make a variation of those packs to let homeowners, businesses, and utilities store the energy they generate during the day so they can use it at night, or in emergency situations, or whenever the price to draw power from the grid is especially high. These packs are called Powerwalls. A Powerwall is about the size of a fridge door and flat enough to mount on a wall. Musk says a ten kilowatt-hour Powerwall will be available in various colors for $3,500 apiece. (Tesla would later produce a fourteen kilowatt-hour version that retailed at $5,500.) With battery solutions like the Powerwall, we can better take advantage of the world's number one power source, which is abundant, cheap, and shouldn't exhaust itself for another five billion years or so. "We have this handy fusion reactor in the sky called the sun," Musk says to laughter from his fans. "You don't have to do anything—it just works. Shows up every day and produces ridiculous amounts of power."

With the combination of solar panels and good batteries, it is possible to transition the whole world from fossil fuels to sustainable energy, Musk has calculated. He shows us the math. A picture of the Tesla Powerpack appears on-screen. It's a white metal box that looks like a sumo wrestler's coffin. It contains a tall stack of Tesla-made battery packs. All we need is two billion of these Powerpacks and we could do away with fossil fuels altogether. Two billion, Musk contends, isn't all that many. That's about the number of cars and trucks on the road—and those fleets get refreshed every couple of decades. "This is actually within the power of humanity to do," Musk says with subdued insistence. "We have done things like this before."

The crowd remains quiet, as crowds that have been confronted with math are wont to do.

Musk says Tesla will make its first Powerpacks and Powerwalls at its factory in Fremont, but will ramp up production when it moves into the Gigafactory. Then comes a surprise: Tesla is planning to build multiple gigafactories. "There's going to need to be many other companies building sort of Gigafactory-class operations of their own, and we hope they do," he says. Tesla will make its Gigafactory patents available to use free of charge, just as it does with its vehicles. The crowd applauds. He describes the Gigafactory as a "giant machine" and asks people to think of it as a product. It just happens to be a product that's stuck in the ground.

Tesla's founding goal was to prove that electric cars could be better than gasoline cars, and now it wants to get people excited about a clean-energy future. "We want to show people, most importantly, that this is possible," Musk says. The combination of solar power, batteries, and electric cars is the only available option that he knows can halt the deadly progress of the Keeling Curve.

In November 2016, Tesla demonstrated the potential of what's to come by announcing the details of a clean-energy project on the island of Ta'u in American Samoa. Tesla had combined 5,328 solar panels with sixty Powerpacks to cover almost all of the island's energy needs for its six hundred inhabitants. The island previously relied on diesel generators for power. In January 2017, Tesla revealed a much larger installation in Southern California, where four hundred of its Powerpacks would offset the loss of the Aliso Canyon natural-gas storage facility. The battery packs could store enough energy to power 2,500 homes for a day. In July 2017, the company agreed to build a 100-megawatt storage system in South Australia—three times the size of the next-largest lithium-ion storage system in the world. That October, Tesla installed a makeshift battery plant to supply power to a children's hospital in Puerto Rico, which was reeling without power and basic amenities in the wake of two devastating hurricanes.

Tesla is what the entrepreneur and venture capitalist Chris Dixon of Andreessen Horowitz calls a "full-stack start-up." The word *stack* here is borrowed from computer programming. It refers to the group of things that make up a product or service, its support mechanisms and applications. If you own the "full stack," you have holistic control of the system. In this way, philosophically and strategically, Tesla bears more in common with Apple than it does with GM or Toyota. Just like Apple, Tesla wants to control the whole experience around its business, from the design of its battery packs, to the making of the software, to the manufacturing of vehicles, the building of components, the management of infrastructure, and the sale of its products through its website and retail stores. The first Gigafactory is the essential structure underlying this entire system: producer of the fundamental building blocks of Musk's full-stack vision.

This vision came into full view on July 20, 2016, when Musk published a follow-up to his 2006 blog post about Tesla's "secret master plan." Ten years on, much of that plan, which laid out the company's premium-to-mass-market strategy, had been realized, save a promise to "provide zero emission electric power generation options." His update to the vision was headlined MASTER PLAN, PART DEUX, a comedic reference to the sequel to the *Hot Shots!* movie. The plan came a month after Tesla had made its offer to acquire SolarCity, the solar energy provider run by Musk's cousins Lyndon and Peter Rive and for which Musk served as chairman.

In "Part Deux," Musk again emphasized that the world is facing a fossil-fuels-induced crisis, and the faster the world moves to sustainable energy, the better. Tesla had changed the mission statement on its website from "accelerate the world's transition to sustainable transport" to "accelerate the world's transition to sustainable energy." The CEO's language in the blog post doubled down on that shift, adding strength to a claim he had made at the launch of the Powerwall: Tesla was now an energy innovation company. On a press call after announcing Tesla's offer to acquire SolarCity, Musk had said, "The world does not lack for automotive companies. The world lacks for sustainable energy companies." He would also claim that combining Tesla and SolarCity created the potential for "a trillion-dollar market cap company"—the most valuable ever known.

The next phase of Tesla would focus on four areas, Musk said in his blog post. It would pair its Powerwalls with SolarCity's energy business to create an integrated generation-and-storage product "that just works." Family homes could be their own utilities, and customers would get "one ordering experience, one installation, one service contact, one phone app." In the years ahead, Tesla would also expand its vehicle fleet, adding a compact SUV, a pickup truck, a heavy-duty truck, and a small bus into the mix. The buses would be autonomous, to be summoned by smartphone app, or via buttons at existing stops. The advent of full self-driving capability, which Musk said would ultimately be safer than human-driven vehicles by an order of magnitude, would also enable a business built around car-sharing. Owners could add their cars to Tesla's shared fleet to generate income when they weren't using them. In cities where there weren't enough customer-owned cars to meet the demand for such shared-use cases, Tesla would operate its own fleet—a move that would put it in direct competition with Lyft and Uber.

Of course, it's easy to write down a wish list of things for your company to do, and it's far from easy to actually achieve those things. Musk was criticized for failing to offer substance. A *Bloomberg* columnist called it "less like a plan, more like a manifesto," and Reuters said it was "big on vision, short on detail." These shots missed the point. Musk knows that storytelling has power, and that the mere act of articulating goals matters a lot to their prospects of eventual achievement. In 2006, the Model S was just a line in a blog post—"the second model will be a sporty four door family car"—and not even the Roadster was available for sale. In 2013, the Hyperloop was just one man's science project, sketched out through rough calculations in a white paper that he had written in an all-nighter. In the same month, the Gigafactory was just an uttered thought, when Musk wondered aloud about the prospect of building a "gargantuan factory of mind-boggling size."

For now, at least the Gigafactory part of Musk's new story is a reality. While Tesla decided to build the factory because it was essential to meet the production needs of the Model 3, it has since said that as much as half of its output will be dedicated to energy storage. It is thus the bedrock of two potentially giant businesses—or could be. Even in its advanced state of progress, the first Gigafactory project is so complex, enormous, and expensive that it embodies daring and danger in equal proportions. At millions of square feet and a cost of $5 billion, its scale is heart-stopping. Even with a $1.6 billion contribution from Panasonic, Tesla has had to raise billions for its construction through bond sales, share sales, and lines of credit. And there's no guarantee the costs won't inflate before the Gigafactory's planned completion in 2020. Any setbacks that delay its completion wouldn't be limited to the facility itself. If the Gigafactory falters, Tesla's overall business will falter. It's not just the Model 3 but also the energy storage business. It's the future of the Tesla pickup, the heavy-duty Tesla Semi, the automated fleet. Musk's full-stack vision is nothing without the Gigafactory at its base.

But this time around, Tesla's potential competitors are not waiting to see how the crazy bet works out. While traditional automakers have (on the whole) not yet committed to building their own battery factories, other groups are moving to occupy positions as strong as Tesla's—just not in America. In August 2017, a German consortium called TerraE announced that it would start building a 34 gigawatt-hour lithium-ion battery factory in late 2019, with the goal of reaching full capacity in 2028. At the same time, a range of smaller companies in China plan to build enough factories to account for more than 120 gigawatt-hours of capacity by 2021.

With the Gigafactory, Tesla has again positioned itself as a pioneer in an industry worth many billions of dollars. Anyone interested in the future of transportation and the energy economy should watch that building down the road from USA Tavern with utmost attention.

⁓⁓⁓⁓⁓

It was close to 100 degrees Fahrenheit at midday on Tuesday, July 26, 2016. The wind was whipping up loose dirt at the Gigafactory site, so the workers and construction equipment were intermittently enveloped in mini tornadoes. There was no urban pollution—no soda cans, no plastic bags, and no cigarette butts gathering in gutters. The nearest neighbors in the vast industrial park were distribution centers for PetSmart and Toys "R" Us. The land, largely barren, was dotted with sagebrush, and a ring of mountains, snowcapped in the winter, encircled the building site. The men and machines toiled in a giant depression in the earth, the floor pressed flat, with plots marked off to indicate the sites of future sections of the mammoth factory taking shape around them. Bulldozers, graders, and pickup trucks scurried like hyperactive ants amid shipping containers, cement foundations, and mounds of reddish dirt. A thousand construction workers were on the site seven days a week, rushing to meet the accelerated timeline set by Tesla in the wake of the unexpected bonanza of orders for the Model 3.

Phil LeBeau, to whom Musk had first publicly mentioned the possibility of building such a factory three years earlier, was on the site. Wearing a hard hat and goggles, he did a live cross to the CNBC studio from inside the partially completed facility. "This is Tesla's Gigafactory, your first look inside," LeBeau intoned. The floors were gray and shiny, and huge pillars supported the high ceiling. In the background, a mess of rolling tables and shelves waited to be put into order. Bright LED lights hung from overhead lamps. "One-point-nine million square feet have been built out at this facility—that's only fourteen percent of the total manufacturing footprint." The space behind him was already large enough to house an indoor amusement park, and there would be three stories of this—four in some parts. The walls were temporary to facilitate further expansion.

Ultimately, the building would stretch a quarter of a mile long and cover 5.8 million square feet in floor area, sitting on 3,200 acres of land. There was room enough for another right next to it. Musk wanted it to produce up to 150 gigawatt-hours by 2020, enough to support 1.2 million cars. That would likely mean employing more people than the original estimate of 6,500. As many as 10,000 jobs could be created here within a four-year time span. LeBeau tweeted a photo of a scale model of the factory that showed solar panels on the rooftop and robots like the ones in the Fremont factory occupying enclosed rooms for assembly work, standing by beltlines and containers for stacks of battery packs.

The building, already enormous, sat heavy on the earth. It would grow into the shape of a diamond—"aligned true north," Musk had promised—in the years ahead. Its squat body was mostly white—a signature color for Tesla's production facilities—with a ribbon of red on its shoulder and a strip of gray at its feet. Half of the structure that had been erected was covered and enclosed. The rest lay open, an exposed steel skeleton ready for concrete. The rooftop was bare, not yet adorned with solar panels, waiting like a vast, empty swimming pool. From the sky, it looked like a chip on a circuit board, the factory's future outline enclosed by a rectangle of road, with parked cars, stacked steel, and etchings in the dirt as the transistors and resistors.

In an afternoon press conference held in the lobby in the factory's northwest corner, Musk would further underscore the importance of the Gigafactory. It wasn't just about the cars. He said Tesla's energy storage business would be as big as its car business in the long term. He said that the company would probably even get into grid services eventually. No doubt that would add more rivals to its list—not just Big Oil and Big Auto but the electric power industry, too. As Musk expands the scope of Tesla's ambitions, he simultaneously ensures that his company will have further political fights ahead. He's looking to take the Gigafactory international, too. There'll be one in Europe, one in China, and probably one in India, he said.

Musk lives among movie stars in Los Angeles and commutes to work among the inventors of Silicon Valley, but he is best suited to the Wild West. Here, there is nothing but a place to build. At the press conference, perched on a stool with microphone in hand, he enthused about the romanticism of his diamond in the desert and noted that there were ten thousand wild mustangs in the area. The horses, present in the United States since they were brought over by the conquistador Hernán Cortés in the sixteenth century, were protected under a 1971 act that deemed them "living symbols of the historic and pioneer spirit of the West." In the Gigafactory, on lands inhospitable to humans, the horses had an unusual new neighbor. But there was at least some mutual benefit. To aid its construction efforts, Tesla had created a series of ponds, and the horses had developed a habit of hanging out nearby. It was a convenient arrangement. When thirsty, the horses simply led themselves to the water.

Three days later, Tesla celebrated the official Gigafactory opening with a boisterous party for more than two thousand employees, owners, officials, fans, and the press. Musk and Straubel kicked off the party with a speech and slide show in a tent erected for the occasion. Fun facts were noted: The Gigafactory is large enough to house ninety-three 747 airplanes, or fifty billion hamsters. At the end of the speech, the cofounders took questions from the audience. Musk and Straubel revealed that a minibus would be built on the Model X platform, a next-generation Roadster would have to wait until later, and Tesla would be recycling batteries at the Gigafactory. Then, in one of the final questions of the night, a man shouted from the crowd: "How can we help?"

Musk didn't hesitate.

"I mean, I know you guys think global warming is real," he said, close to laughter, "but the crazy thing is, a lot of people out there don't. It blows my mind." He wanted his followers to spread the word. "There's a nonstop propaganda campaign from the fossil fuel industry. They're just defending themselves. It's kind of what you would expect"—he shrugged—"but they just, it's nonstop—and they have, like, a thousand times more money than we do."

The partygoers booed the absent foes. Musk urged them to fight back against the messages that muddied the science of climate change and complicated the advent of a sustainable energy future. "The revolution's going to come from the people."

12

WE DIDN'T RUN OUT
OF STONES, EITHER

"The revolution is not a molecule. It's the system."

Joshua Brown's Tesla Model S was traveling eastward at seventy-four miles per hour along the US 27 Alternate, a two-lane highway outside Williston, Florida, in clear, sunny weather on May 7, 2016. Brown, on his way back from a trip to Orlando's Walt Disney World, was in the right lane and, given what happened next, did not seem to be paying attention to the road ahead of him. He had activated his car's Autopilot, Tesla's advanced driver-assistance software, which meant the car was designed to keep itself in the lane, maintain a steady speed, and, in an emergency, brake or swerve to avoid a collision.

Brown was a forty-year-old ex–Navy SEAL who lived alone in Canton, Ohio. During the US war in Iraq, he had helped the navy dismantle bombs. Upon returning to Ohio, he started a company that provided Internet service to people in rural areas. He loved his Tesla, gave it the nickname "Tessy," and, in his first nine months of ownership, drove it more than forty-five thousand miles. A month before the trip to Disney World, he had posted a video to YouTube that showed dashcam footage of his Tesla, in Autopilot mode, automatically swerving to avoid a collision with a truck that was cutting into his lane. Musk had tweeted the video, garnering more than 2,400 retweets and helping to bring it to the attention of the news media. "@elonmusk noticed my video!" Brown tweeted at the time. "I'm in 7th heaven!" Now that Musk had noticed something of his, Brown told a friend at the time, he could die and go to heaven.

Eight days after the tweet, Musk said the "probability of having an accident is fifty percent lower if you have Autopilot on," before adding: "Even with our first version, it's almost twice as good as a person." Tesla's cars carried warnings to drivers that the system was in beta. When Autopilot was activated, a message box popped up on the digital dashboard: "Always keep your hands on the wheel. Be prepared to take over at any time."

As Brown's Tesla approached an intersection on that Florida highway, he did not appear to notice that an eighteen-wheel truck hauling a white trailer had cut off the road in front of him. The truck, which had been traveling in the opposite direction, had not completed a left turn into a side road when the Tesla, holding its line and without slowing down, slammed into the underside of the trailer. The impact tore the roof off the car and sent it careening off the highway. It plowed through two fences and smashed into a power pole, which sent it spinning for a half turn before finally coming to rest hundreds of yards from the intersection and just a few feet from the front door of a family home. Brown's Tesla looked like it had been stomped on by a mastodon. The sixty-two-year-old truck driver, who managed to keep his vehicle on the road, was uninjured. Brown was killed on impact.

The public did not learn of the accident until fifty-five days later, when Tesla published a blog post disclosing that the National Highway Traffic Safety Administration (NHTSA) was opening a preliminary investigation into the crash. In the post, Tesla defended the safety of the system and noted that this was "the first known fatality in just over 130 million miles where Autopilot was activated."

The accident was widely considered the first death in a "self-driving car"—though that term overstated the scope of the technology. As Tesla defended itself, a flood of speculation-filled stories variously attempted to explain why Autopilot wasn't to blame—or why it was— and examined, based on little more than a cursory police report, to what degree each of the two men involved were culpable. Some opinion writers wondered if the crash would set back the cause of autonomous driving, while others contended that it proved nothing. As is often the case, the chief effect of what must have been a painful public "litigation" for the truck driver and Brown's family was to oversimplify a complex situation that was being investigated by better-qualified and less biased parties—in this case, NHTSA.

But Tesla's two-month delay in publicly releasing information about the crash raised some questions. Reporter Carol Loomis asked why Tesla had not revealed details of the accident earlier. Loomis, who after sixty years working at *Fortune* was one of the country's most celebrated financial journalists, had spent her July Fourth holiday asking Tesla why it had not disclosed the fatality ahead of a stock sale.

"On May 18, eleven days after Brown died, Tesla and CEO Elon Musk, in combination (roughly three parts Tesla, one part Musk), sold more than $2 billion of Tesla stock in a public offering at a price of $215 per share—and did it without ever having released a word about the crash."

Loomis thought that Tesla might be in breach of US Securities and Exchange Commission (SEC) rules.

"To put things baldly, Tesla and Musk did not disclose the very material fact that a man had died while using an autopilot technology that Tesla had marketed vigorously as safe and important to its customers."

Responding to Loomis's request for comment via e-mail, Musk argued that half a million people a year would have been saved worldwide if Tesla's Autopilot were universally available. "Please, take 5 mins and do the bloody math before you write an article that misleads the public," he wrote; the episode was "not material to the value of Tesla."

Again, here was a complex situation being examined with blunt instruments. On one hand, Loomis and *Fortune* concluded that the nondisclosure was "very material." On the other, Musk and, later, his supporters, contended that it was not. Loomis noted that Tesla's stock price had initially fallen from $212 to $206 the morning after Tesla's blog post was published. By the end of the day, however, the stock had climbed above $216. It is fair to assume that, on July 1, 2016, there were factors other than the Autopilot accident that may have affected Tesla's stock price, but that's not the way the episode played out in the media. (Ultimately, it would be up to neither Musk nor Loomis to decide.) *Fortune* stayed on the case, publishing a follow-up story that noted Tesla had said in an SEC filing, updated on May 10, that "failures of new technologies that we are pioneering, including autopilot in our vehicles," could lead to product liability claims.

On the same day, electric car news site *Electrek*—which has taken a Tesla-tinted view of the world so often that it sometimes crosses the line into advocacy—published a piece that said *Fortune* was twisting Tesla's words. The language from the SEC filing was a "boilerplate risk disclosure statement" that had existed for at least two years. Also, *Fortune*'s report had omitted the part of the statement that noted that the auto industry experiences significant product liability claims and that Tesla would be at risk of exposure to claims "in the event our vehicles do not perform as expected resulting in personal injury or death." *Electrek* concluded that Autopilot is not meant to prevent an accident like the one that killed Joshua Brown.

The news site summed up its version of events by suggesting that

Fortune was a beneficiary of a multimillion-dollar public relations offensive against electric vehicles, funded by billionaire industrialists Charles and David Koch, whose Koch Industries is one of the largest private companies in the United States. Koch Industries and its subsidiaries own major oil operations, including refineries, pipelines, and leases on at least 1.1 million acres of tar sands in Canada. *Electrek* pointed out that *Fortune* had earlier published an opinion piece by Koch Industries board member James Mahoney. In that piece, Mahoney argued that electric cars shouldn't get government subsidies. "The serious bottom line here is that *Fortune* is getting Koch money to further their climate changing agenda," *Electrek* wrote at the time.

If *Electrek* were in the mood for conspiracy, it might also have mentioned that Carol Loomis is a close friend and bridge-playing buddy of Warren Buffett. Loomis has for many years edited Buffett's shareholder letters. In her article, she should have disclosed her relationship with Buffett, who at the time was entangled in a dispute with Musk and SolarCity over solar subsidies in Nevada. Buffett's investment company, Berkshire Hathaway, owns NV Energy, a public utility that generates most of its electricity from natural gas, and was lobbying against solar subsidies and policies that helped SolarCity and threatened its own bottom line.

Electrek's accusation wasn't quite on the mark. The Mahoney piece that *Fortune* published on its website was part of a pageview-driven play for free content called the Fortune Insiders network. Fortune Insiders was similar to a scheme from competing publication *Forbes*, which offers contributors space on its website to publish articles at no cost. Such moves had become common among publications looking for ways to increase their pageview counts so they could attract advertisers who pay per digital impression. Publications such as *BuzzFeed*, *The Guardian*, and the *Huffington Post*, among many others, have their own spaces set aside for such content. Unfortunately, these "contributor networks" have the run-on effect of muddying the distinction between content produced and edited by the publication's paid staff and that contributed for free by motivated outsiders, many of whom are not governed by the ethics of professional journalism. *Fortune* promoted its Insiders section as "an exclusive group of contributors who share their ideas with our global online audience," but it reviewed the content only for "grammar, clarity, and taste." Mahoney's piece was thus not a pay-to-play effort by the Kochs but mere opportunistic exploitation of a modern media gray area that can leave some readers confused as to who stands behind what.

For Musk, however, *Electrek*'s suggestion that *Fortune* was directly benefiting from Koch money was convincing enough. He tweeted a link to the *Electrek* story that made the accusation and added two words: "Sponsored articles . . ."

As evidenced by his warning at the Gigafactory opening party that the industry was putting out "nonstop propaganda," Musk was already primed to defend against oil industry attacks on his businesses. At a conference in May 2013, he was asked if he had a message for the nation's oil companies. Musk replied that it's hard to ask oil companies to act against their best interests. The incentive structure of the prevailing system, he noted, applied no penalty to dumping carbon dioxide into the oceans and atmosphere. He said he was in favor of a carbon tax that would encourage better behavior, just as cigarettes and alcohol are heavily taxed because of their bad health effects.

A couple of minutes later, after a digression about the necessity of taxes, Musk did raise one complaint. "I guess where I have an issue with the oil and gas guys is where they sometimes engage in nefarious tactics," he said, "or things that are somewhat insidious, like funding academic studies that people can then point to as though they have some credibility, and it's some prominent professor somewhere—but that person has been paid off by the oil industry to write that study." Such behavior should be condemned in the strongest terms, he said. He recommended that people read *Merchants of Doubt*, by science historians Naomi Oreskes and Erik Conway. "That actually spells it out in detail how some of these things are going on, where the oil and gas industry, all they need to do is create doubt—and that's what they've done." In fact, some oil groups had employed the same individuals and firms that the tobacco industry used to sow doubt about the link between smoking and lung cancer. "I'm surprised that some of these people are still around because, like, you know, they're quite old," he said, with a wry laugh.

In *Merchants of Doubt*, which was published in 2010, Oreskes and Conway show how special-interest groups conducted sophisticated public relations campaigns to prevent government action on a range of environmental and health issues tied to industry and commerce. The groups realized that they didn't need to disprove findings that worked against their interests, but that creating doubt was enough to confuse the public and get politicians to assist in their efforts to slow down the regulatory process. These individuals and groups repeatedly used such tactics to stymie public-minded measures to mitigate or eliminate the effects of smoking, acid rain, the ozone hole, and DDT, to name a few. The same playbook is being used against climate science today, and since 2010's *Citizens United v. FEC* Supreme Court ruling, which made it possible for interest groups to spend unlimited amounts of money on political activities without public disclosure, the practice has become even more opaque.

Attacks on climate science in recent years have included a sustained character assault on Dr. Michael Mann, a climate scientist from Pennsylvania State University whose research showed a sharp and sudden rise in temperatures from the 1850s on, compared to a gentle decline in average temperatures over the preceding 850 years. When graphed, the temperature averages over the course of a thousand years produced a shape like a hockey stick—the long, constant line of slight decline was the stick's shaft, while the near-vertical line since the Industrial Revolution was the blade. Although Mann's research would later be backed up by more than a dozen separate studies, he and his work were called into question by groups and individuals such as the fossil-fuels-funded George C. Marshall Institute and Fred Singer, both of whom had instrumental roles in a sustained campaign to downplay the links between secondhand smoke and lung cancer. Mann's work was also attacked by Willie Soon, a part-time Smithsonian employee with a PhD in aerospace engineering often misidentified as an astrophysicist who, it was later revealed, accepted more than $1.2 million from the fossil fuels industry over the course of a decade, including at least $230,000 from the Charles G. Koch Charitable Foundation. Joining the detractors was Senator James Inhofe, a Republican from Oklahoma who, according to OpenSecrets.org, had received more than $100,000 of campaign donations from Koch Industries from 1989 to 2016. Inhofe has repeatedly described global warming as a "hoax."

This doubt-creation engine has also been central to the New York attorney general's fraud investigation into ExxonMobil over the gap between what the company has known for decades about climate change science and the public relations efforts it allegedly used to create the impression that the science was inconclusive. For instance, Exxon helped create a fossil-fuels-industry lobby group called the Global Climate Coalition, which peddled the idea that the role of carbon emissions in climate change was "not well understood." Exxon's own scientists had repeatedly provided evidence that climate change was a serious concern and could have harmful consequences. Exxon has denied any deception, alleged a conspiracy among its antagonists, and responded to the investigation by claiming its free-speech rights were under attack.

The ongoing climate-science denial campaigns are so numerous and pervasive that they must get their full airing in other books. But it is enough to know that they are real, they happen, and the effort to undermine such science has become an industry unto itself, involving a network of fossil-fuels-funded politicians, academics, media, think tanks, nonprofits, and other groups that purport to be acting in the interests of the free market—a subject dear to the Kochs' hearts. The Kochs believe that the role of government should be not just limited but almost nonexistent. They advocate the privatization of social security, the eradication of subsidies, weaker labor laws, and minimal regulation of the environment, to name a few favorite issues. Since the passage of Citizens United, it has become more difficult to track spending by big-money donors like the Kochs, but a 2013 Drexel University study found that conservative foundations had distributed as much as $7 billion to climate-denier organizations from 2003 to 2010. The billionaire donors behind the foundations had largely been able to conceal their involvement.

It is not without reason, then, that Musk was concerned about the potential for the Koch brothers to be colluding with *Fortune*. Indeed, in February 2016, the journalist Peter Stone reported that a Koch-backed group was being assembled to spend $10 million a year to attack electric vehicle subsidies. The effort, Stone said, was being led by James Mahoney and Charlie Drevna, a lobbyist and member of the Koch-funded think tank the Institute for Energy Research and, until 2015, president of the American Fuel and Petrochemical Manufacturers— a position previously held by Mahoney.

If you're wondering why the Kochs might have been worried about electric cars when the cars represented less than 1 percent of all vehicles on the road, you might consider an ionic compound called sodium chloride. Oil may have dominated our economic times, but it fades in significance when compared to salt. Civilization was built on the stuff that McDonald's now gives its customers for free.

Salt is essential to the human diet—without it, your body would gradually shed water in an attempt to maintain constant salt levels in your blood, eventually leading to your death from thirst. But it has been almost as important as a means to preserve food. Over the centuries, humans have harvested salt in a number of ways, from mining to evaporation to digging up bogs that had been soaked with seawater. Such methods have been traced back at least 3,500 years, with evidence that they stretch back even 5,000 years, and some bear similarities with how oil is extracted today. In AD 400, the Chinese discovered a way to drill into mountains and extract brine with bamboo pipes, some of which reached as deep as three thousand feet.

Salt, like oil, was unevenly distributed around the world. Thriving settlements arose around salt sources in Jordan by the Dead Sea; in North Africa, where salt could be dug from the ground; in the Austrian Alps, where salt was mined; and in Persia, Egypt, and the Sahara, where there were salt swamps in the deserts. In parts of Africa where salt was scarce, people got their salt hits by drinking the blood and urine of cattle and wild animals. It was the world's most important commodity and so became the subject of transport, trade, and conflict.

"A certain political pattern seems to emerge," wrote the journalist M. R. Bloch in *Scientific American* in 1963. "Where salt was plentiful, the society tended to be free, independent, and democratic; where it was scarce, he who controlled the salt controlled the people." In the civilizations of the Nile, Babylon, India, China, Mexico, and Peru, autocratic rulers controlled their subjects by maintaining a monopoly on salt, and taxing it.

While today's global economy remains inextricably linked with the fortunes of the oil industry, salt's connection to the economy was even more direct. Salt was synonymous with money, and in some cases literally was money. Ethiopia used bars of salt as currency as early as the sixteenth century and, in remote areas, as recently as the twentieth. The word *salary* has its origins in the Latin word for "salt money." The Romans paid their civil servants in salt. Slave traders bought humans with it.

There were, of course, wars. In Roman times, German tribes fought over salt sources. France's salt tax, the *gabelle*, caused such outrage that it was an aggravating factor leading to the French Revolution. Even during the American Civil War, salt was a military target. At the end of 1864, for instance, Union forces captured Saltville, Virginia, a leading producer of the stuff, then embarked on a destructive two-day rampage that, according to the historian Rick Beard, effectively brought an end to salt making in the South of the United States.

These days, dieticians say our problem is too much salt, not too little. So if salt carried such strategic importance only 150 years ago, why is it so cheap today? The answer is that it was supplanted by an invention that changed the course of history.

The first refrigerated ships appeared in the mid-1870s, and General Electric started marketing the first household refrigerator in 1911. Instead of relying on salt for food preservation, or using large ice blocks to keep their food cold, people in developed countries started storing it in electrically chilled boxes. Food became safer, lasted longer, and tasted better. This revolutionary development facilitated the rise of large modern cities, the opening of global markets for food, and the spread of population. It also made salt a lot less valuable. There would be no more wars over sodium chloride. Its multithousand-year reign as the world's most important commodity was over.

What happened with salt is not that it was displaced by a superior ionic compound. It was displaced by a superior system. The same thing is happening with oil.

The oil industry may be the most lucrative the world has ever known, and the idea that still-scarce electric cars pose a serious imminent threat to it might seem fanciful. The industry is worth trillions of dollars a year. The production, supply, and distribution of oil is the subject and cause of much geopolitical instability, and it has been central to conflicts on every continent, from the Middle East to Sudan and the South China Sea. While it continues to be fought over, and while the burning of oil continues to warm the Earth's atmosphere in an unsustainable way, it's also important to acknowledge that oil, like salt, has been essential to the vitality of modern society. The United States of America as we know it would scarcely hold together without an abundant supply of gasoline to fuel the cars and trucks that connect its highly dispersed towns, cities, and agricultural areas. We still depend on oil to maintain our quality of life, to enjoy freedom of travel, and to connect global economies. If oil disappeared immediately, life for many would quickly become grim.

None of that, however, means that oil is not vulnerable to the same forces that made packaged salt a free item at fast-food outlets. In 2014, about 47 percent of petroleum products consumed in the United States were used for gasoline, according to the US Energy Information Administration (EIA). While oil is used for many other products, such as jet fuel, plastics, and detergents, the wealth of the industry is fundamentally dependent on cars, trucks, and buses. Without gasoline and diesel filling the tanks of motor vehicles, the oil giants of today would be far less significant players in the global economy.

It doesn't take much to trigger an oil market crisis. From June 2014 to January 2015, an oversupply of oil sent prices crashing from $116 a barrel to $47 a barrel, prompting an industry panic. Oil companies big and small laid off staff and canceled hundreds of billions of dollars of projects. Supply had been driven up by a number of factors, including the shale boom, which, in 2012 and 2013, resulted in the fastest growth in United States oil production history. The improving fuel efficiency of America's vehicle fleet also contributed. According to the EIA, the US transportation system used 10 percent less oil in 2014 than it did in 2007. As electric cars become more widespread, the demand for oil will further decrease, putting more pressure on oil prices and creating more economic stress in the industry. Shell has said that oil demand could peak as soon as 2021.

The displacement of two million barrels of oil a day—about 2 percent of global daily production—would be enough to trigger oil price decreases equivalent to those seen at the start of the crisis in 2014, according to a story by *Bloomberg* that drew on a 2016 study by Bloomberg New Energy Finance. Electric cars could do that by the early 2020s, the study found. The growth rate of electric cars from 2014 to 2015 was 60 percent, which was similar to the growth rate Tesla was projecting for the years ahead. If that rate continued, electric cars could displace two million barrels of oil a day by 2023, *Bloomberg* noted. A more conservative estimate based on the component costs of electric vehicles and when they would be affordable to mainstream car buyers found that the two-million-barrel threshold would be crossed in 2028.

But might that time come even sooner? As discussed, both Tesla and GM think battery prices will come down fast enough for electric cars to be more affordable than equivalent gasoline cars by the early 2020s. The Chevy Bolt sells for less than $35,000, after subsidies. Tesla plans to be producing Model 3s at a rate of hundreds of thousands a year by 2019. Other electric car companies, new and old, are developing competitive strategies.

It is still difficult to predict how quickly the sales of electric cars will overtake those of gasoline vehicles. Even assuming all goes well for Tesla and their electric competitors, it could take years, or decades. Bloomberg New Energy Finance's study estimated that electric cars will account for 35 percent of new car sales by 2040. That's based on battery prices decreasing at a slower rate than Tesla and GM anticipate. But, as noted earlier, gasoline cars will face the difficult task of competing with electric cars that are both cheaper and better.

One characteristic of disruptive technologies, as the electric car has the potential to be, is that their market penetration tends to start slowly and then accelerate rapidly. In 1900, less than 10 percent of US households had access to electricity. In 1960, less than 10 percent of US households owned a color TV. In 1990, less than 10 percent of US households had a cell phone. The first versions of all these products tended to be expensive, clunky, inconvenient, or all of the above. But then, as the technology improved, manufacturing processes were refined, and economies of scale kicked in, prices came down dramatically and the technologies found their way into homes and pockets. In 1990, there were 5.3 million cell phone subscribers in the United States—about 2 percent of the population. Twenty-five years later, 92 percent of Americans owned a cell phone. When mapped on a graph, this adoption curve looks roughly like a stretched S—a gentle incline at first, followed by an inflection point that triggers a sudden and steep rise, and then, ultimately, a leveling off when the technology reaches saturation point. Over the last hundred years in the United States, the "S-curve" has occurred with the automobile, the radio, the color TV, the microwave, the VCR, the personal computer, the cell phone, and the Internet. Oh, and the refrigerator.

Could the electric car follow the same path? Count Elon Musk among the believers. "At the beginning of last year [2015], we had fifty thousand cars in total on the roads worldwide, and then last year we produced another fifty thousand cars," he said in January 2016. "So the total fleet of Tesla vehicles doubled last year. It will approximately double again this year."

We shouldn't take Musk's word for it, of course—Tesla's 2016 production fell about twenty-five thousand cars short of doubling the previous year's tally—but consider that many of the effects that spur demand for electric vehicles are only just starting to take hold. The decline of battery prices, which will make electric cars more affordable, is probably the biggest factor influencing demand, but there are others. For a start, many hundreds of millions of people still don't know a thing about electric vehicles that aren't golf carts or hybrids like the Toyota Prius. They might be unaware of the benefits of instant torque, or the near-total silence of the propulsion, or that the vehicles can be charged at any power point. Tesla, with its fancy stores, slick websites, and high media profile, has captured a hard-core loyal market, but there's a lot more market to be had.

Traditional automakers spend billions of dollars a year on advertising to encourage people to buy their products. In 2013, GM alone spent $5.5 billion on advertising. Tesla, on the other hand, has spent virtually nothing on advertising its cars. Automakers invest in advertising because it correlates with increased demand. What will happen once Tesla and others start paying to advertise the benefits of electric mobility?

But no matter how much you spend on ads, if customers can't get near the cars, they won't buy them. If you live in New Zealand, for instance, you weren't able to buy a Tesla through official channels until 2017. Many cities in the United States don't have a Tesla store, and most Americans haven't sat in a Model S or Model X—or any other electric car. As more full-electric cars get on the road, more people will be able to experience what they're like and realize that they're much different from golf carts and Priuses. Tesla has long believed that the best way to sell its cars is to get people in them. Once a potential customer has taken a Tesla for a test drive, she is more likely to buy one. Many Nissan Leaf owners say they'll never go back to gasoline cars.

And then there's the wild card of regulation. Market forces already suggest that electric cars will soon be more affordable than gasoline cars independent of rebates and other incentives, but even slow-moving governments with conservative expectations for how quickly things can change are considering regulatory packages that seek to end the sale of gasoline cars within two decades. Every country in the United Nations has committed to drastically reducing its carbon emissions, and automakers have been expected to continue to improve the fuel economy of their vehicles (although President Donald Trump has withdrawn the United States from the Paris climate pact). But the global political environment could get even worse for gasoline cars if the effects of climate change wreak more havoc with the world's economy and way of life—particularly if affordable, low-emissions alternatives are readily available. For example, Norway is working on a combination of taxes, subsidies, infrastructure, and other incentives in an effort to end sales of gasoline cars in the country by 2025. In October 2016, Germany's federal council voted for a nonbinding resolution to end all sales of gasoline cars with internal combustion engines by 2030. In May 2017, India's power minister announced a plan to have only electric cars—and "not a single petrol or diesel car"—sold in the country from 2030 on. Both the UK and France have said they will end sales of diesel and gasoline cars by 2040. And even China has said it will set a date that will signal the end of all gasoline car sales in the country (although it hasn't said what that date will be).

All these scenarios could have a drastic effect on the uptake of electric vehicles, which would in turn have a dramatic impact on the consumption of oil. Even by Bloomberg New Energy Finance's relatively conservative estimates, there will be enough electric cars on the road to cause an oil crash in the late 2020s. And every year from then on, the story will only get worse for the oil companies. Bloomberg's study forecast that electric vehicle sales will leap from 462,000 in 2015 to forty-one million in 2040. Every new electric car on the road represents another dent in the oil companies' profits.

"It's clear that the sector is going through one of the most transformative periods in its history, which will ultimately redefine the energy business as we know it," said a PricewaterhouseCoopers report on oil and gas trends in 2016. But it's not just industry job losses, writedowns, and budget cuts that will come. Geopolitical power structures will be rewritten, from oil-rich regions in the Middle East and Africa to oil-import-dependent nations elsewhere. National security priorities will shift.

Saudi Arabia, which has traditionally relied on the petroleum sector for 90 percent of its state budget, is responding. Prince Mohammed bin Salman, next in line to the throne, has control over Saudi Aramco (Saudi's oil monopoly), economic policy, and the national investment fund. He has announced plans to create a $2 trillion fund to make returns from investments, not oil, the primary source of Saudi government revenue.

If Saudi Arabia, the world's largest exporter of oil, is concerned about the S-curve, you can see why the Koch brothers might aggressively defend their oil-related business interests and so keep that profit channel open as long as possible. Yet that aggressive defense can serve as further reason to believe that oil will go the same way as salt.

I met Rich Sears at the Tresidder student union building on Stanford University's campus in Palo Alto, three miles from Tesla's headquarters. It was a bright and still summer afternoon, so we decided to sit outside. We picked a round steel table underneath a tree in the courtyard. Sears worked part-time at Stanford as a consulting professor in the Department of Energy Resources Engineering. A geophysicist, he had worked for Shell for three decades, rising to the level of vice president, and dedicated much of his time to oil exploration. While still on Shell's payroll, he spent eight years as a visiting scientist at the Massachusetts Institute of Technology and, after retiring, went on to be a senior advisor on the commission that investigated the 2010 BP Deepwater Horizon oil spill in the Gulf of Mexico—the worst offshore environmental disaster ever.

I first came across Sears, a tall, lean man with a long nose and deep-set eyes, when I watched a TED talk he gave in 2010 about

planning for the end of oil. In that speech, he noted that there were a
hundred trillion gallons of crude oil in the world still to be developed
and that it would never run out. "It's not because we have a lot of it,"
he said. "It's not because we're going to build a bajillion windmills. It's
because, well, thousands of years ago, people had ideas—innovations,
technology—and the Stone Age ended. Not because we ran out of
stones." Sears was paraphrasing former Saudi Arabian oil minister
Sheikh Ahmed Zaki Yamani, who made the statement in 2000. Innovation would provide the way out of the oil era.

Sears's career has taken him all around the world, but he now lives
in a house on a hill in Danville, California, with his wife. He owns a
1952 MG in fire-engine red, a machine so beautiful that he custom-
built a clapboard garage especially for it. Inside, he has hung old Shell
memorabilia on the walls. That day at Stanford, Sears was wearing the
relaxed uniform of Silicon Valley—a polo shirt and jeans. He spoke
slowly and somewhat theatrically, peppering his speech with *guess
whats* and *by the ways*, in the manner of someone accustomed to making a case before a crowd.

As he said in his TED talk, Sears believes that it's technology that
drives great economic change, and that the same will prove true for
the world's energy economy. "The revolution is not a molecule," he told
me. "It's the system."

It was Sears who gave me the idea that oil could be thought of in
the same way as salt. "Right now," he said, "when people talk about
the end of oil, lots of people will quickly go to 'Okay, well, what's it
going to be instead?' and they start thinking about 'Well, what molecule can I make that you would use instead of gasoline in your internal combustion engine?'" But that's not what killed salt. "Salt didn't
die as a global and strategic commodity because somebody found
another molecule that they could go dig up in the dirt that was better." The real culprit was refrigeration, which seemingly came out of

nowhere. "Who would have thought?" Sears asked, his tone rising in feigned disbelief.

Sears was sympathetic to the idea that electric vehicles might bring about the end of oil, especially when car-sharing services, like Uber, were added to the equation. Toward the end of our interview, we started discussing the factors that might help or hinder the widespread adoption of electric vehicles. The subject of regulation came up, and he posited that, if it looked like millions of jobs would be lost in the disruption of, say, Detroit's auto industry, politics might slow the rate of progress.

I suggested to Sears that the Koch brothers would fund obfuscation efforts against electric cars, too, and interfere with the regulatory process.

The Kochs have an extensive history of using their money and power to influence politics and government. They are strong opponents of government regulation and subsidies, particularly as they pertain to environmental policy. They have also demonstrated an ability to obstruct regulation that would address climate change even through market-based means, such as a carbon tax. Looking at Koch Industries' businesses and past troubles with the regulatory sector, it's not hard to see why the brothers—each worth about $50 billion and among the top ten richest people in the world—might object to government action on carbon emissions and other pollution. For example, the Kochs own more acres of Canada's tar sands than any other non-Canadian company, including Exxon, Chevron, and Conoco.

While Koch Industries and its subsidiaries have a diverse set of business interests, from polymers and fibers to forestry and cattle ranches, they also preside over a fossil fuels empire of infrastructure, refineries, pipelines, storage, and shipping, and they profit from fossil-fuels-related financial instruments, such as an oil derivative that they co-invented. The University of Massachusetts Amherst's Toxic 100 Air

Polluters Index ranked Koch Industries eighth in the United States for
toxic air pollution and noted that it released 29.4 million pounds of
toxic chemicals into the air in 2014. Koch Industries has responded by
noting that the Amherst index includes "virtually every manufacturer
in the United States today," and that one of its cocreators, Michael
Ash, is a member of the Union for Radical Political Economics, an
organization of academics and activists that critiques capitalism. In
2014, Koch Industries was a bigger polluter than Valero, Chevron, and
Shell. In 2000, it paid a record $30 million to settle suits with the EPA
for serial environmental crimes, including more than three hundred
oil spills in six states. That same year, subsidiary Koch Petroleum agreed
to spend $80 million to reduce refinery emissions and pay a $4.5 million penalty to the EPA.

And in 2009, another Koch subsidiary, Invista, paid a $1.7 million
penalty and promised to spend half a billion dollars to fix more than
680 violations of EPA standards at its facilities. Koch Industries has
claimed that it has a positive relationship with the EPA and that it has
received hundreds of environmental, health, and safety awards since
2009. It has said it is continuing its efforts to improve its environmental performance. All that is publicly known about Koch Industries'
financial standing is that it brings in about $115 billion in annual
revenue, according to a 2014 estimate.

Both David and Charles Koch are prolific charitable donors, particularly in cancer research and the arts, but their spending on political influence has taken on history-making proportions. David Koch
embarked on an unsuccessful campaign as the vice presidential nominee for the Libertarian Party in the 1980 US presidential election,
running to the right of Ronald Reagan, but the Kochs have since
stayed away from podiums and focused instead on funding research,
candidates, and organizations that promote their view of economic
freedom.

Over the course of four decades, they have spent many millions of dollars to fund think tanks, academic institutions, donor groups, public relations campaigns, and politicians that support their cause. By 2015, the brothers had established a political network of several hundred wealthy donors—many of whom are coal, oil, and gas magnates—staffed by 1,200 employees in 107 offices around the country. The organization was three and a half times larger than the Republican National Committee, according to an analysis by *Politico*, which called it a "private political machine without precedent." The Kochs' network pledged to spend $889 million on the 2016 US election, more than either the Democratic or Republican Party. (However, after Donald Trump, whose populist economic views conflicted with their own, became the Republican nominee, the Kochs decided to scale back their spending to $750 million.) *The New Yorker*'s Jane Mayer has described these efforts as a "40-year project that Charles and David Koch have been funding with their vast fortunes to try to change the way Americans think." The journalist and climate activist Bill McKibben has said the Kochs "may be the most important unelected political figures in American history."

Back on the Stanford campus, as our conversation turned to how regulation could affect the transition to electric vehicles, Sears questioned the wisdom of focusing on the Kochs. He wondered whether or not a community of climate scientists could do a better job of communicating and building a coalition to change people's minds about climate change action "rather than whine about the Koch brothers." He felt the Kochs were a "little blip on the radar."

Sears was leaning back in his chair. It was close to 5:00 P.M. and we had been speaking for more than two hours. He had brought a bottle of Diet Coke to our meeting. It now lay empty on its side on the table in front of us. Summer school students were bringing out pitchers of beer to the tables around us.

Sears believed that people on all sides of the arguments about climate change were guilty of distorting science. "If I happen to be an academic and I want to do climate studies, and my life and future is about getting support for my academic studies," Sears said, "there's a lot of money out there to support a certain point of view."

I agreed, and he continued. "Well, am I being cynical? You know, in the same way that the Koch brothers are trying to preserve their world, I think this is an equal-opportunity sport. There are a lot of people trying to preserve their world."

I replied that the economic incentives for each side were lopsided. "There's probably some scientists who are making their research fit neatly with the accepted view so that they can ensure they've got a safe career," I said. "But I think that that's pretty small compared to a multitrillion-dollar oil industry that stands to lose a lot more if they're forced to meet these regulations."

In May 2016, there appeared a website called WhoIsElonMusk .com. Most prominent on the site was an autoplaying video on the home page that featured foreboding music and the dark, smoky-voiced narration of a true-crime TV show. It spent two minutes shifting through B-roll footage and clips pilfered from documentaries about Musk. The video announced itself with the title *American Swindler: The Elon Musk Story* and carried an ominous-sounding warning:

> Foreign-born billionaire Elon Musk. His companies are synonymous with technology and wealth, and his jet-setting lifestyle is the envy of the world. But how exactly did Musk's companies come about? Who has Elon Musk exploited along the way? And whose world is he actually changing? The truth may startle you.

The video proceeded to suggest that Musk has been using his "unprecedented access to the halls of power" to line the pockets of politicians in an effort to secure billions of dollars of subsidies for his ventures Tesla, SpaceX, and SolarCity—all at the expense of the unsuspecting American taxpayer. Below the video, the site linked to critical articles about Musk and his "crony capitalism," including opinion pieces written by such people as Veronique de Rugy, a research fellow at the Koch-funded Mercatus Center at George Mason University, and Bruce Fein, formerly an adjunct scholar at the American Enterprise Institute, funded by the Koch-linked Donors Trust nonprofit group. In his piece, Fein even wondered if the Koch brothers should create "an annual Elon Musk Fleecing Government Award to stigmatize wealth begotten from the hijacking of government to obtain risk-free riches."

When Keith Cowing, author of the blog *NASA Watch*, discovered the site, he grew suspicious and looked to see who owned the domain. By looking around in the source code, he found a different URL that was registered under the name Brad Summey, the chief technology officer for a political ad agency named Orange Hat, based in the outskirts of Washington, DC. According to OpenSecrets.org, Orange Hat had previously been paid $220,200 by supporter groups for Minnesota Republican representatives John Kline and Erik Paulsen, both repeat beneficiaries of funding from Koch Industries and its associated political action groups. For the 114th Congress (2015–2016), the Kochs' political advocacy arm, Americans for Prosperity, gave Kline and Paulsen scores of 87 percent and 91 percent respectively. The scores indicate the extent to which the representatives' votes in Congress align with the lobby group's view of economic freedom.

There was a group that officially claimed responsibility for WhoIsElonMusk.com—it identified itself as the Center for Business and Responsible Government, although no such group exists in official records.* It claimed to be "a non-partisan organization dedicated

---

* A search did return a result for "Center for Responsible Business and Government," which correlates with the e-mail address listed on the website's domain registration. A limited liability company of that name is registered in Delaware, but it returns no other results in a Google search.

to highlighting cronyism and its effect on American taxpayers and policy." Although there is no proven link between the center and the Kochs, it is a common Koch tactic to fund political front groups to do their bidding under generic or even virtuous-sounding names, such as Citizens for a Sound Economy, Citizens for the Environment, and Center for Individual Rights. In some cases, such groups are no more than a checkbook, some contractors, and a lockbox.

The *American Swindler* video carried a caption that referred to $4.9 billion in subsidies that Musk has received for his industries, with a credit to the *Los Angeles Times*. Indeed, in May 2015, the *Times* carried a piece by Jerry Hirsch that reported Musk's companies had collectively benefited from $4.9 billion in government subsidies, according to data compiled by the newspaper. The article counted the zero-emission credits trading scheme that Tesla has benefited from and construction-related incentives from Nevada, Texas, and New York for Tesla, SpaceX, and SolarCity, such as the billion-dollar incentives package afforded to Tesla for the Gigafactory. The article quoted hedge fund manager Mark Spiegel, who short-sells Tesla stock, as saying Musk's companies wouldn't be around without government support.

Three days later, the *Times* published Musk's response. He called the subsidies "a pittance" compared to government support for fossil fuels companies. The International Energy Agency has estimated that the fossil fuels industry collects about $550 billion in global government subsidies annually, compared with about $120 billion for the much smaller renewables sector. The United States has given oil companies tax breaks worth more than $470 billion over the past century, dating back to 1916, according to data compiled by *Mother Jones*. Supporters of renewables, meanwhile, argue that anything that combats climate change should of course be subsidized. "If I cared about subsidies, I would have entered the oil and gas industry," Musk told the *Times*.

A reasonable market-based analysis of the subsidies received by fossil fuel companies should also count their effective gain from not having to pay for the mess their products' emissions have made of the atmosphere. A mom-and-pop store, for instance, cannot just leave its trash strewn on the sidewalk. The owners have to pack it up neatly and pay, either directly or through taxes, for garbage collection. Fossil fuel companies have never had to pay for their garbage collection. They've been free to pump carbon emissions into the atmosphere for more than a hundred years (of course, when they started, no one knew just how pernicious this trash was). Two weeks before the *Times* printed its article about Musk's subsidies, the International Monetary Fund (IMF) published a report that put the fossil fuel industry's garbage bill for 2015 alone at $5.3 trillion—a figure that took pollution and climate change into account.

In May 2016, at the World Energy Innovation Forum, hosted at Tesla's Fremont factory, Musk said, without citing evidence or names, that oil industry representatives had been "shopping around" the *Times* story to journalists in the wake of the IMF report. "They got the *LA Times* to bite on a bullshit story that was totally nonsensical," he said. He said it didn't make sense to compare subsidies that Tesla would receive over a twenty-year period for the Gigafactory with fossil-fuel industry subsidies that were a thousand times larger in a single year.

In November 2016, another anti-Musk website appeared as part of a campaign to call into question the subsidies that Tesla has received. The site, StopElonFromFailingAgain.com, was backed by the political action group Citizens for the Republic, chaired by Laura Ingraham, a conservative radio host and advocate for Donald Trump's presidential campaign. (Ingraham has decried former Democratic senator Harry Reid's criticism of the Koch brothers as a "disgusting" demonization.) Citizens for the Republic also had on its board a man named Craig Shirley, a one-time lobbyist for Citizens for State Power, which fought regulation of utilities at the turn of the century. Citizens for State Power was later revealed to be secretly funded by utility groups.

Koch Industries has embarked on a campaign to say that it's not against electric cars; it just wants the government to stop subsidizing them because that favors one form of energy over another. In August 2016, Charlie Drevna announced that the advocacy group he had established with James Mahoney, called Fueling U.S. Forward, was not actually about attacking electric cars but was instead focused on promoting the positives of fossil fuels. In a Koch Industries advertorial published in a special edition of US politics newspaper *The Hill* in April 2016, the company wrote that fossil fuels and electric vehicles can and should coexist "on a level playing field." The language in the advertorial closely mirrored the language in James Mahoney's opinion piece that was published on *Fortune*'s website. Both articles called out the Department of Energy (DOE) loans program that benefited Tesla as well as solar panels manufacturer Solyndra, which went bankrupt in 2011 and became a favorite political punching bag.

In his 2016 testimony to a House of Representatives subcommittee on energy, Nicolas Loris decried the market "distortion" of the government's involvement in such deals as the DOE loans. Loris worked for the Heritage Foundation, which had received funding from the Charles G. Koch Charitable Foundation, where Loris previously worked as an associate. In November 2014, the DOE announced that the renewable-energy loans program was profitable and later revealed it would net $6 billion once all loans had been repaid. The loss on Solyndra, which amounted to $528 million, made up the bulk of the portfolio's losses.

Congressmen who have received campaign donations from the Kochs have also proven willing to apply pressure to Tesla specifically. South Dakota Republican senator John Thune, who in 2016 was chairman of the Senate Committee on Commerce, Science, and Transportation, called on Tesla to explain to the committee its response to the Autopilot fatality. Thune's letter to Tesla found its way to Reuters before it was delivered to the company. Thune, who said that "manufacturers must educate consumers not only about their benefits but also their limitations," has accepted more than $50,000 in campaign contributions from Koch Industries in his career, according to OpenSecrets.org. Thune was running for reelection at the time he called Tesla to task.

There's also a connection between the Kochs' favored politicians and an SEC investigation into Tesla. Senator Mike Crapo, an Idaho Republican also running for reelection in 2016, served as the chairman of the subcommittee that oversees the SEC. Crapo scores a lifetime grade of 93 percent on the Americans for Prosperity scorecard. Between 2011 and 2016, he accepted campaign contributions of $40,000 from Koch Industries (the National Automobile Dealers Association is another of his major donors). It may be a coincidence—and there is no evidence to suggest otherwise—but on July 11, 2016, a source close to the SEC revealed to *The Wall Street Journal* that the commission was investigating whether or not Tesla breached securities laws by failing to disclose the Autopilot fatality ahead of its stock sale—a question first raised by Carol Loomis at *Fortune* a week earlier. Again, the media knew about the government action before Tesla did. The SEC's prompt action against Tesla contrasted with its reluctance to investigate ExxonMobil over its alleged failure to disclose to shareholders possible material risks due to climate change. The SEC had ignored an October 2015 letter from four members of Congress who were supportive of climate-change action, urging it to investigate Exxon. At the time, the attorney general of New York was investigating Exxon for evidence of fraud (since the state's public pension funds are Exxon shareholders), but the SEC chose not to act.

Back at Stanford, Rich Sears made the case that people aren't seriously willing to sacrifice quality of life in exchange for giving up fossil fuels—"We're not going to all freeze in the dark"—so environmental activists should stop demonizing the oil companies and the Kochs. "You still need them."

This is true. We have all benefited from oil throughout our lives. Much of the food we eat is transported by gasoline-burning vehicles. Almost every time we get in a car, we're moved from place to place by the combustion of gasoline and air. We fly from country to country thanks to oil. The lifestyles we demand are the primary reason why oil companies are so profitable. It would be unfair to condemn them for selling us something that we demand virtually every day.

"Now," Sears continued, "if you want things to change and you recognize that one of the problems is that nobody's really paying for the carbon, fine. Then put a price on carbon. And the oil companies have all said—and they've all been serious, they're not just saying it—put a price on carbon. Just do it."

"So why doesn't that get passed?" I asked.

"I think because of the politics of it, the way it's handled," Sears replied. "The environmental community does itself a terrible disservice when they want to talk about climate deniers and whine about the Koch brothers, and sue Exxon. How can you go suing Exxon because they suppressed climate science? 'They knew in their labs that this was a problem.' Yeah, and how did you find that? Because they published all those papers. They weren't suppressing anything."

Sears was right that the large oil companies agreed to put a price on carbon. Exxon, BP, and Shell had all spoken in favor of it. A tax would ensure "a uniform and predictable cost of carbon," an Exxon spokesperson said, and "allow market forces to drive solutions." It was one thing they could agree on with environmental groups. In June 2016, however, the Republican-controlled House of Representatives approved a resolution to condemn such a tax. Here's how *Bloomberg* explained it:

"The House strategy, pushed by Majority Whip Steve Scalise, a Louisiana Republican, and backed by Koch Industries Inc., used the symbolic measure to lock in votes against a tax on carbon dioxide emissions blamed for climate change. The tactic was designed to weaken the ability of a future president and Congress to levy [a carbon tax] to help pay for a broad overhaul of the US tax code, said Republican strategist Mike McKenna."

Sears was also right about Exxon going public with its climate change studies. The company's scientists had published peer-reviewed papers and spoken at conferences about the effect a warming planet would have on the industry and humanity. However, the criticisms directed at Exxon concerned not the company's secrecy but its role in creating doubt about climate science through public relations campaigns and other public statements.

Sears went on to argue that environmentalists had turned global warming into a belief system, that labeling people "climate deniers" doesn't help anyone change their minds, and it doesn't help bring people with different viewpoints together. But he was not advocating inaction. On the contrary, he believes we should get off oil regardless of whether or not climate change is human-powered, or even a bad thing.

"On general principles, the fact that we're burning all of this legacy carbon, pulling it out of the ground in huge quantities and dumping it into the atmosphere as carbon dioxide, on first principles, that's a bad idea."

Sears shared a key belief with environmentalists: The world needs to stop adding to the concentration of carbon dioxide in the atmosphere—and fast. But he recoiled at the thought of an oversimplified political debate when, as he pointed out, "these are all complex systems."

"The talking heads don't want to acknowledge the complexity," he said. "They want a simple story that boils down nicely to the sound bite, to the headline, and they want to run with it. And so it ends up being the Koch brothers are bad, and climate change is bad, and Exxon is bad—and, you know, I think people ought to be for stuff, not against stuff. I think you could be much more effective if you were for something, rather than being against the Koch brothers."

Then Sears's tone brightened. "Elon Musk—whatever you think of Tesla—Elon Musk didn't build Tesla by whining about the internal combustion engine."

"Right," I agreed.

"He could've!"

"Well, now he is whining about the Kochs," I said, adding: "Quote unquote 'whining.'"

"Okay, fine," Sears said with a sigh. "He's wasting his time. He needs to focus on something more positive. But seriously, what he has done, what he has built, was not built through whining."

"He inspired a vision of hope," I offered.

"That's right. Yeah. And just maybe it will turn into part of an energy transportation something-something revolution. The whining's not going to change any of that. The whining's not going to make the Koch brothers go away. Eventually, they'll die and the government will take most of their money and that will be that."

Yes, eventually the Kochs will die—Charles is in his eighties, and David is in his late seventies. But they may leave legacies greater than funds contributed to government coffers. Blocking the passage of a carbon tax supported by both environmentalists and oil companies is one such legacy. Creating confusion about clean-energy subsidies and slowing the transition to electric mobility could be another.

13

HEAVEN OR HELL?

*"There are certain important things that we must do
in order for the future to be good."*

I met Carsten Breitfeld in July 2016 on the sunny patio at a hotel on San Jose's Santana Row, a glitzy pedestrian shopping mall. We sat under the half shade of an umbrella, amid fountains and hotel guests in tank tops and sunglasses, happily chatting over breakfast. Breitfeld, in a polo shirt and jeans, had brushlike brown hair kept tame on the sides with a buzz cut. It was a Saturday morning and he seemed relaxed, but it was his first day off in six weeks. He lives in Hong Kong now, but he was visiting Silicon Valley on a tour of tech companies with which his new car company, Future Mobility, later renamed Byton, might partner. He had been focusing on artificial intelligence and was encouraged by what he saw. "All the basic pieces are in place," he said, sitting back in his chair with his right foot balanced on his left knee.

Breitfeld, a twenty-year veteran of BMW who left to become CEO of Byton, holds the somewhat contrarian view that Tesla is outdated.

"To some degree, Tesla is some kind of traditional car company already," he told me as he picked at a fingernail. He spoke with caffeinated energy, his German accent moderated by years of international travel. Even the Model 3 left him unmoved. "I do think it's a good car, but it's still a very normal car."

Breitfeld promised that Byton—funded by the Chinese luxury auto retailer China Harmony New Energy Auto, Chinese Internet giant Tencent, electronics manufacturer Foxconn, and retail conglomerate Suning, among others—would produce something much more advanced than any of Tesla's vehicles. What Tesla has done is only the first step, he said. The second step is to convert the car into an intelligent object. Byton's cars will include a digital experience for every passenger and connect with other Internet services to personalize the mobility experience. Your car should know when your next meeting is and prepare itself for your commute. It should know you on a personal level, like a whip-smart chaperone. The key for this? Collecting data on a scale that compares to Google or Apple. "Someone told me, if you don't believe in God, maybe those guys are the ones in the world who know you best," Breitfeld said, referring to the tech behemoths.

Breitfeld joined BMW in 1996 after completing a PhD in mechanical engineering from the University of Hannover. He worked in numerous departments at the company, climbing up along the way: chassis; brakes; slip controls; transmission; driveline; and finally into the executive ranks as head of corporate strategy, power train, and cooperation, reporting to then CEO Norbert Reithofer. While Breitfeld was in that position, the company devised a strategy to reduce carbon emissions with a hybrid electric vehicle. The idea was to develop a sports car of Porsche 911 quality by combining a small three-cylinder combustion engine with a powerful electric motor.

And so was born "Project i," BMW's program for the i8, a luxury plug-in hybrid sports car with a zero-to-sixty-miles-per-hour time of

4.2 seconds. The program would also produce the compact all-electric i3, which could travel 114 miles per charge. Breitfeld, installed as head of the program, was given two orders from the CEO: The i8 had to be completed within three years, and it had to be of outstanding quality.

Breitfeld had to figure out how to work around the company's standard operating procedure and build a high-performance team, which would work at a site independent of BMW headquarters. As the program progressed, an entrepreneurial spirit infused the team. "They gave us this freedom to act, bypassing all the processes. This created a lot of power and energy." But others in the company were dubious. Breitfeld's colleagues told him he was crazy. "What did you do?" they asked, thinking he was being punished for some unknown transgression. "This will never work," others concluded.

After a year, Breitfeld's team had come up with a prototype for the i8. It was ugly, but it worked, with 100 kilowatts of power on the front axle from an electric motor, and 170 kilowatts in the rear from an internal combustion engine. "This prototype, even if it was done very quickly and not refined at all, gave all the people the feeling, wow, this could be something really great," Breitfeld recalled. Others in the company started to take more of an interest.

By 2014, Breitfeld had met his deadline and BMW launched the i8 during a weeklong international press event in Santa Monica. The car, which sported doors that opened almost vertically like the wings of a butterfly, attracted immediate interest. Early reviewers referred to it as a "dream machine" with "equal parts sex appeal and efficiency." Both the i8 and i3 have garnered praise since, despite their relatively high price tags (the former costs about $140,000 and the latter, $43,000) and, by BMW standards, low sales numbers. These days, seemingly everyone at the company claims to have been involved with Project i in some way or another. "This means success!" Breitfeld joked, laughing at the thought.

However, within BMW, the i3 and the i8 were seen as niche

products, secondary to the 3 Series, 5 Series, and 7 Series that constituted the backbone of the company. That's a mentality that pervades
the industry, Breitfeld says. "They're doing too many cars right now
with the old technology, earning a lot of money out of it, being very
profitable," Breitfeld said. "Attacking the car portfolio would mean
that customers would go from the traditional product to the new
product." Unfortunately, the new products cost more to make and
deliver slimmer profits. "They are struggling very much to do this
transition."

The challenge of the technology shift is exacerbated by short-term
thinking, Breitfeld said. Board members at the traditional car companies tend to have three-to-five-year contracts and are often close to
retirement age. They think three years out, not fifteen. "They're con-
centrating very much on today's and tomorrow's businesses."

With the i8, Breitfeld had had a taste of the future and wanted
more. He imagined what a company would look like if it combined
Uber's ride-sharing model with electric propulsion and autonomous
driving. "This would be a machine to print money," he said, breaking
into a smile. Such a vision couldn't be realized at BMW, so he started
considering other options.

He was approached by some Silicon Valley tech companies, including Tesla. The proposition was tempting, but he felt Tesla was
lacking crucial manufacturing expertise. "You need traditional car
people who know how to run a production process."

The meeting that ultimately led him to Byton took place while he
was on vacation at Lake Garda in Italy at the end of 2015. Changge
Feng, chairman of the Harmony group, had contacted Breitfeld to see
if he would be interested in starting a car company. Breitfeld wasn't
about to give up his vacation to go to China, but he told Feng he
would give him ninety minutes over lunch if he was willing to meet
at Lake Garda. Several days later, Feng and three associates arrived at
the marina where Breitfeld's yacht was moored. The investors, armed

with documents, had researched Breitfeld carefully. Feng was empowered to hire him on the spot. All he had to do was sign on the dotted line.

Breitfeld didn't want to move so quickly. Not only was he not entirely convinced by the initial selling of the plan, but he also wasn't excited about the idea of moving to China. The basic story was sexy, Breitfeld said, but it had flaws. Feng seemed in a rush. "You can start today and we will go to market at the end of next year," he told Breitfeld. The German responded that that timeline was impossible. However, he accepted an invitation to spend a long weekend in China for more-in-depth discussions.

Breitfeld and his wife were living in Munich at the time. It's a beautiful city, an easy lifestyle. When Breitfeld told his wife that he was going to China for the meetings, she assented, but made one thing clear: "We are not going to move to China." She changed her mind once she saw Hong Kong, a cosmopolitan city that compared favorably even to Munich.

The discussions in China provided more substance. Breitfeld impressed upon the investors that a car project would take patience and a lot of investment. They would need to spend a billion dollars before they could hope to earn a dollar back. Also, they would have to push back their launch plans. "This is an Internet car," they said, "so we have to move fast."

"Yes," Breitfeld returned, "but it's still a car." There would be tooling, testing, homologation. These things take time.

Ultimately, Breitfeld was pushed over the edge by China itself. He recognized the emergence of an entrepreneurial culture. Hordes of people in their late twenties, full of optimism and brimming with entrepreneurial passion, were starting companies. But more importantly, the political framework was right. "It's the best in the world," Breitfeld said matter-of-factly. The market is large, the middle class is on the rise, and the government is committed to supporting new-energy vehicles. Byton would be selling the government on the idea that autonomous driving adds value to society: It reduces accidents and associated injuries and deaths, and it provides private mobility options for the elderly, among other benefits. The company would set up a testing environment where it could legally demonstrate the technical capabilities of autonomous vehicles. It planned to have a drivable prototype of its first car, a premium SUV with an iPad-like touch screen integrated in the steering wheel, ready by 2018 and a production version on the road in China in 2019. It would have three cars, including a sedan and a multipurpose seven-seater, on the road by 2022. Byton aimed to take customers not as much from fellow electric-car company Tesla as from Germany's premium auto brands BMW, Audi, and Mercedes-Benz. It has announced that it will build the cars in a factory in the eastern Chinese city of Nanjing that it will spend $1.7 billion to build.

I asked Breitfeld if there were other people like him who were still at traditional automakers but wanted to try something new. He pointed out that he had hired the core team from BMW's i8. That group included Dirk Abendroth, who developed the electric power trains; Benoit Jacob, who was head of design; and Henrik Wenders, head of product management. Byton had also hired Luca Delgrossi from Mercedes-Benz to head up its autonomous driving unit, and Mark Duchesne, formerly of Toyota and Tesla, as head of manufacturing. "At the end of the day, it's about the idea of the company and what they really want to do," Breitfeld said. He scratched his head and shifted in his seat. "This industry is transforming to a completely new era. The good ones want to be part of it."

Byton encapsulates the main characteristics of the modern multinational auto start-up. It has offices in China and Silicon Valley and is funded by Chinese financiers. It boasts German engineering leadership and American design expertise. Its vision is for connected, autonomous cars that have business applications beyond just unit sales, and it sees Tesla as a trailblazer whose cars will nevertheless come to be seen as passé. Like Nio, Faraday Future, and Lucid Motors, Byton has the braggadocio of youth and the optimism that comes with a fresh start. However, it also shares the same vulnerabilities.

Trying to get too big too fast created huge problems for Jia Yueting's LeEco, throwing the fragile empire into question. It became apparent that things were going badly for Jia in late 2016, when he sent a letter to his employees to confess that LeEco was running out of money. A few months earlier, it had signed a deal to buy American TV brand Vizio for $2 billion, just the latest in a series of expensive endeavors over the preceding year and a half (the deal ultimately fell through because of what Vizio called "regulatory headwinds"). LeEco had also purchased a 70 percent stake in the ride-sharing company Yidao Yongche ($700 million), acquired 29 percent of smartphone maker Coolpad ($450 million), and bought new-media broadcasting rights to Chinese Super League soccer (close to $400 million). Jia had personally invested $300 million in Faraday Future and put up $5 billion of LeEco stock as collateral for tax abatements for its factory in Nevada.

The week after Jia sent his letter, Faraday ceased construction of its factory and ultimately, in July 2017, abandoned it completely. The financial future of Lucid Motors was also cast into doubt, since at least 45 percent of the company was directly or indirectly funded by Jia and his interests. (In 2017, the company was working hard to raise money from other sources.) In his letter to employees, Jia also singled out LeEco's car division, LeSee, for profligate spending. LeEco pulled out of its deal with Aston Martin, which scaled back its plans for the RapidE, pushed its launch date back to 2019, and instead announced its intention to produce a line of electric cars based on its Lagonda model.

"No company has had such an experience, a simultaneous time in ice and fire," Jia wrote in his letter. "We blindly sped ahead, and our cash demand ballooned." Jia promised that the company would forge ahead "in disruption and pain." A week later, LeEco announced $600 million of extra funding, and in January 2017 it confirmed a lifesaving $2.18 billion investment from Sunac, a property development giant from Jia's home province, Shanxi. LeEco had also broken ground for a vehicle assembly plant in Zhejiang province, on China's east coast. The company said the plant's capacity would accommodate the production of four hundred thousand cars by 2018.

Faraday Future, however, was not a direct beneficiary of LeEco's funding injection. As the company prepared its first vehicle for production, it was in dire straits. A series of news reports in 2017 revealed that it owed debtors hundreds of millions of dollars, and it was on the receiving end of lawsuits that complained of missed payments. In July 2016, outside accountants told Faraday's executives that the company had misjudged the extent of its liabilities. In fact, while Faraday thought it had $100 million on its balance sheet, it actually owed $200 million—a $300 million discrepancy. Its seat supplier, Futuris, sued for missed payments totaling more than $10 million, but the suit was dismissed for undisclosed reasons at an early stage. Beim Maple Properties, the owner of a warehouse rented by Faraday, filed a suit claiming that the company had missed more than $100,000 in rent payments. *BuzzFeed* also uncovered an October 10 letter from construction company AECOM to Faraday warning that it was $21 million behind on payments for its factory and that another $37 million would be owed for work done in October and November.

There was an exodus of executives. In the second half of 2016, six high-ranking employees left Faraday in the space of a few months. Included among the departures were general counsel James Chen (who had joined from Tesla), finance director David Wisnieski, operations controller Syed Rahman, and head of product strategy Robert Filipovic. In December 2016, two more executives—chief brand and commercial officer Marco Mattiacci, formerly the president and CEO of Ferrari North America and Ferrari Asia Pacific, and vice president for product marketing and growth Joerg Sommer, a former executive at Volkswagen, Daimler, Opel, and Renault—left the company. Even more followed, including founding executive and vice president of human resources, Alan Cherry; chief financial officer Stefan Krause, who had joined from BMW and Deutsche Bank; chief technology officer Ulrich Kranz, formerly of BMW; and head of manufacturing Bill Strickland, who had led the development of the Ford Fusion. The latter three have since started their own electric car company, Evelozcity, and have been joined by Richard Kim, former head of design at Faraday. Ultimately, in 2018, founders Tony Nie and Nick Sampson left too.

Perhaps hoping to overcome the negative press, Jia and Faraday staged a much-hyped presentation at the Consumer Electronics Show on January 4, 2017, to unveil the FF 91, its first vehicle intended for production. Unfortunately, the most notable part of the unveiling was that the car failed to drive to center stage and park itself as promised. In driving demonstrations outside the venue, the car worked just fine, but the public embarrassment was a lot to bear for a company that was sagging under the considerable weight of credibility issues.

Despite the setbacks, Faraday remained defiant. In the wake of CES 2017, it claimed to have received more than sixty-four thousand reservations for the FF 91—although it later admitted that only "priority reservations" required a deposit. Anyone under any name could place a "standard" reservation for free. Just over two weeks after the public self-parking failure, Jia tweeted a photo of his presentation with the message: "We believe an easy road may never lead to greatness. Adversity shapes character." He accompanied the statement with the hashtags #Allin and #DreamOn. Nevertheless, Faraday would scale back its ambitions, aiming to produce two vehicle models instead of seven and reducing its initial production targets from 150,000 cars a year to 10,000. After walking away from its plans to build in Nevada, Faraday instead opted to lease a ready-made factory in Hanford, California.

Meanwhile, things went from bad to worse for Jia himself. In July 2017, a Shanghai court froze his assets after a LeEco affiliate missed loan payments. That December, China put him on a blacklist of debt defaulters after he failed to pay Ping An Securities Group about $72 million as ordered by a Beijing court.

Observing these proceedings, Byton might have taken some comfort from having a diversity of funding sources. It had also, by and large, kept a lid on hype. An arena spectacle to sing the praises of a yet-to-be-made fantasy car seemed like the last thing Carsten Breitfeld would want. But that doesn't make the company, or its peers, immune to one of the other major challenges that these hybrid Chinese-American start-ups face.

In any company, the corporate culture is crucial to success. It's difficult to overstate the extent to which steady management, a clear vision, and the right balance between productivity and happiness matter to a company. A 2013 survey of more than 2,200 executives by strategic consultancy Booz & Company found that 60 percent believed culture was more important to a company than strategy or operating model. Eighty-six percent of C-suite executives said their organization's culture was critical to business success, but 96 percent said some form of culture change was needed within their organization. The survey's results suggest that culture issues present a challenge for any company—but they are exaggerated when the disparate cultures of two countries are interwoven. For Byton, Nio, Faraday Future, and Lucid Motors, it must be difficult to blend the Chinese way of business with the Western way.

In China, companies are largely hierarchical, with most power flowing to the bosses. Underlings are discouraged from questioning authority and are expected to work long hours on rigid schedules, six days a week. There is little separation between work and personal life. In the United States, by contrast—and particularly in the technology industry—there are more fluid dynamics of authority, with employees at all levels empowered to make decisions and act on them quickly. Creativity tends to be favored over discipline, and forward progress is made through trial and error. In China, the education system relies heavily on rote memorization and the passing of strenuous exams where depth of knowledge is prized above all else. In the United States, a more liberal education system prevails, in which students are expected to be resourceful, think critically, and solve problems through action. These principles flow into the workplace.

"There are always clashes between the Americans and the Chinese," Shaun Rein, managing director of the China Market Research Group, told me. Chinese CEOs are used to micromanaging everything, Rein said, which often doesn't sit well with people within most American organizations. "If it's a new company and the guy is putting his billions of dollars into it and starts to put his name in, he starts to become micro-controlling." Relatedly, China also has a shortage of mid-level executives. China's economy has developed rapidly in recent decades, and from a small base, so founders tend to be young—the average age of billionaires in China is fifty-three years old—and people have been becoming executives in their late twenties and early thirties. "You're missing an entire generation of elder business statesmen," Rein said. When Chinese companies make acquisitions abroad, they are buying not only a brand but also management know-how. The chasm between the micromanaging CEO and the lower-level executives can create problems.

Indeed, according to numerous former employees I spoke to, there have been culture clashes at Faraday Future and Lucid Motors, particularly over questions of authority and who gets to set the strategy. Recall that Bernard Tse, Lucid's founding CEO, left the company after major shareholder Beijing Auto attempted to force it to focus primarily on China, and that former Tesla CEO Martin Eberhard left Lucid after just six weeks on the job because it was being run "like an old-school Hong Kong company." Recall, also, the stories of

American executives at Faraday's Los Angeles headquarters who felt all the decisions were being made in China.

There are still risks even to those new car companies that can either overcome the international divides or don't have to deal with them. Marrying a technology industry mind-set with the traditional mind-set of the auto industry will continue to be tough, not just for incumbents such as BMW, GM, and Ford but also for newcomers such as Che He Jia, Singulato Motors, and Nio. Tech people want to move fast and be bold. Auto people want to move deliberately and according to a tried-and-tested playbook. Only Tesla has proven an ability to take a Silicon Valley approach to automaking and turn it into a money-making enterprise—and even then, it has much to prove when it comes to long-term sustainability.

Greater challenges will come, both known and unknown. In the latter category is the uncertain geopolitical climate. Much of the world is in political turmoil, and the consequent effects on the economy, industry, and climate are still playing out. In 2016, Donald Trump, a political novice with unclear views on climate change and whose own businesses have been bankrupted many times over, was elected leader of (arguably) the world's most powerful country. Dozens of congressmen, advisors, and staffers with close ties to the Koch brothers have been appointed to senior roles in the Trump administration, starting with Vice President Mike Pence and including Secretary of State Mike Pompeo, Former White House counsel Don McGahn, and Scott Pruitt, the former administrator of the Environmental Protection Agency, who has said he doesn't believe that carbon dioxide has been a primary contributor to climate change. Their influence was not significantly offset by one of President Trump's early advisors: Elon Musk. Musk told reporters that he originally agreed to advise Trump because "the more voices of reason that the president hears, the better," but he resigned the position after Trump pledged to withdraw the United States from the Paris climate agreement. The Paris agreement and the country's environmental policies are among a few things dear to New Auto that have been upended by the Trump presidency.

Also in 2016, the United Kingdom, at the time the world's fifth-largest economy, chose to sever its ties with the European Union, raising difficult questions for automakers. They have been attempting to assess how the move will affect supply chain, manufacturing operations, and sales, in the short and long term. The increasing economic instability worldwide leaves open the possibility of another wide-sweeping recession (or worse), an event that would threaten not only the viability of the new car companies but also the traditional automakers, many of which came perilously close to extinction in 2008.

And what of the standard-bearer for the electric revolution?

At a July 28, 2017, event to deliver thirty production versions of the Model 3 to the first owners, Musk gave voice to the question that was on every Tesla critic's mind. "The thing that's going to be the major challenge over the next six to nine months is, how do we build a huge number of cars?" Tesla was making the Model S and Model X at a rate of two thousand a week and had put a total of more than two hundred thousand cars on the road, but it was still a niche automaker. To reach mass-market scale, it would have to produce many more cars and much faster, ultimately pumping out half a million Model 3s a year. By the end of 2017, it hoped to make five thousand cars a week—more than double its midyear production rate—and it hoped to hit ten thousand a week at some point in 2018. "Frankly, we're going to be in production hell," Musk said, before laughing and extending his palms to the thousands of Tesla employees in the crowd. "Welcome, welcome, to production hell."

It was a term Musk had used before and it referred to something that the company's detractors felt could be its downfall. Tesla might be able to make a small number of expensive cars, the argument goes, but it can't be a serious challenger to the traditional auto giants until it can handle the production of millions of cars a year. Such critics haven't had to look hard to find evidence to support their case. For instance, by Musk's own admission, difficulties with the Model X caused a production shortfall in the first half of 2016. Owners had reported a series of issues with the SUV, relating to panel alignments, faulty seats, and, especially, trouble with the falcon-wing doors, which in some cases were found to open at random times or not close properly. Musk confessed to "hubris in adding far too much new technology" to the car. He later said he had been spending nights at the factory to personally oversee quality control. "Basically, we were in production hell for the first six months of the year," he said.

At the Model 3 delivery event, he emphasized that the new car was engineered to be easy to build. Keeping it simple, Tesla initially promised only two variants: a $35,000 Standard version with 220 miles of range, and a $44,000 Long Range option that would drive 310 miles per charge. But, indeed, Tesla faced early challenges in manufacturing the Model 3. In August 2017, Musk had confidently predicted the company would be producing Model 3s by the tens of thousands per week by the end of 2018, but a bottleneck in battery module production from the Gigafactory led him to temper that goal considerably. Instead, Tesla set its sights on a five-thousand-per-week production rate by the end of the first quarter in 2018 (later adjusted to the end of the second quarter). Musk said that Model 3 reservation holders should "assume the worst" in terms of possible delivery dates for their vehicles.

Mass-market manufacturing, however, is just one of many challenges Tesla must overcome to find long-term success. It is a company that continually pushes itself to the limit, trying to take on with thirty thousand employees what many much larger companies would never consider attempting. After all, Tesla's ambition doesn't stop at the Model 3. It's also building a global network of Superchargers, several massive battery factories, and a retail network that reaches to all parts of the planet. Its SolarCity acquisition brought not only financial complications—the younger company had lost more than $758 million in the first three quarters of 2016 when Tesla offered to buy it for $2.6 billion—but also a range of new products to learn and sell, from solar panels to solar shingles. At the same time, Musk has said Tesla will make semitrucks, pickup trucks, minibuses, a next-generation Roadster, and a crossover version of the Model 3, known as the Model Y, all while developing its autonomous driving technology. Tesla is trying to do all this, remember, while not being profitable.

Profitability could be a sticking point for a while. Musk seems much more interested in going for a big long-term win than in chasing short-term profits to mollify shareholders. Tesla continues to invest heavily in massive projects, including the Gigafactory, Model 3 development, its semitruck, and energy storage systems, and it keeps going back to investors and lenders in search of more capital to fund its ambitious expansion plans. The results sometimes make financially minded observers nervous, even if those anxieties aren't always reflected in the stock price. Jim Chanos, a short-seller famous for betting against Enron just ahead of its collapse, has said that Tesla is headed for a "brick wall" and that it will struggle to appeal to customers now that traditional automakers are also making sexy electric cars. "Tesla's biggest asset is its stock price," he said. "When it falls, it will really fall." In late 2017, UBS calculated that if Tesla continued to burn cash at its then current rate—more than a billion dollars a quarter—it would run out of money in 2018.

And yet, Tesla has repeatedly demonstrated an ability to raise funds through creative means, such as accepting preorder deposits for its trucks ($5,000 for a reservation) and the new Roadsters, for which the company asks buyers of the Founders Series to front with $250,000. Buyers of the standard version of the new Roadster, which Tesla says will accelerate from zero to sixty miles per hour in under two seconds, have to pay $50,000 up front. These deposits are worth hundreds of

millions of dollars to Tesla, but it is likely that the company will also
need to raise money through other means, such as a stock sale or further debt offerings, before it can get to profitability. If Model 3 sales
are slow or the economy takes a turn for the worse, the company
could find itself in serious financial trouble. But even then, there's a
fair chance that it won't be too difficult for Musk, who has a legion of
fans in finance, tech, and beyond, to find willing financiers to keep
the car company in business.

Personnel issues are another of the threats that Tesla's critics often
cite. While Musk has defended Tesla's record of retaining executives,
the company has seen many senior employees come and go, and some
stays have been alarmingly short. In January 2015, *The Wall Street
Journal* published a story that said Tesla had suffered from growing
pains because of Musk's "domineering presence." Citing dozens of
interviews with current and former employees, the article pointed to
examples of high-level managers who quit or were fired after clashing
with Musk. "I don't like to fire people. I hate it," Musk told the newspaper. "The issue I've had is firing people too late, not too early." In
March 2017, *Bloomberg* published a similar story, noting that Jason
Wheeler, formerly of Google, had occupied the chief financial officer
role at Tesla for only fifteen months. The story said that there had
been more than two dozen management departures over the previous
twelve months across almost every area of the company. It quoted
unnamed sources as saying that long hours and a tense culture contributed to the departures. Of particular interest was turnover in the
company's autonomous driving team, which, by June 2017, had lost
two key people in six months: Sterling Anderson, who left to start his
own self-driving car company, and Chris Lattner, formerly of Apple,
who came to the mutual agreement with Tesla that he wasn't a "good
fit." Now that traditional automakers are investing more heavily in
electric cars and new entrants like vacuum cleaner–maker Dyson and
India's Tata Motors are aggressively pushing into the market, Tesla

will have to work hard to attract and retain the talent it needs to stay ahead of the competition.

And that's to say nothing of the most important worker at the company: Musk himself. It is impossible not to wonder about his stores of energy and ability to focus on the momentous challenges ahead. Not only is he calling the shots at Tesla, but he's also running SpaceX, a $20 billion enterprise with more than a few ambitions of its own, which include sending astronauts to the International Space Station, a space Internet subdivision, driving the development of cheap reusable rockets, and, ultimately, colonizing Mars. As if he were somehow bored by this trifling workload, Musk has also taken on a host of other side projects, such as Neuralink, a brain-computer interface start-up he cofounded, the Boring Company, which plans to make tunnels for cars, and the Hyperloop, another of his pet interests. Can he do it all?

The job juggling certainly comes with pressures. On July 30, 2017, Musk published a series of tweets that almost amounted to a psychological confessional. Responding to a fan who tweeted that Musk's Instagram account showed an "amazing life," the CEO wrote: "The reality is great highs, terrible lows and unrelenting stress." A separate Twitter user asked if Musk was bipolar. "Yeah," Musk replied. Then he added detail with more tweets:

July 30, 2017, 10:39 A.M.: "Maybe not medically tho. Dunno. Bad feelings correlate to bad events, so maybe real problem is getting carried away in what I sign up for."

July 30, 2017, 10:40 A.M.: "If you buy a ticket to hell, it isn't fair to blame hell . . ."

July 30, 2017, 10:58 A.M.: "I'm sure there are better answers than what I do, which is just take the pain and make sure you really care about what you're doing."

Another indication of the emotional toll of these endeavors was captured in a 2015 interview on Danish television. "Were you a little naive when you thought, 'I can easily build an electric car and a rocket?'" the interviewer asked Musk.

"I didn't think it'd be easy," Musk replied. "I guess that I thought they'd probably fail. But you know, like, creating a company is almost like having a child. So it's sort of like, how do you say your child should not have food?" His eyes took on a sheen.

The interviewer continued. "So once you have the company, you have to feed it and nurse it and take care of it—even if it ruins you?"

"Yeah," Musk replied. His lips started quivering and his eyes got wet. His chin crinkled.

When the interviewer asked how he got through the tough period in 2008, Musk released a heavy sigh and asked if they could take a break. But the camera stayed on. On the verge of tears, Musk shook his head, blinked hard, and looked away.

Musk's comparison of his companies to his children took on new significance for me on July 5, 2017, when my wife gave birth to our first child. On that day, I discovered a new emotional depth and, like many parents these days, worried about the world we were bringing him into. Twenty sixteen was the planet's hottest year on record. Before that, it was 2015. Before that, it was 2014. We have ourselves to blame. The carbon dioxide released into the atmosphere from the fossil fuels we've been burning, largely to power our cars, has compromised our children's well-being. The planet's prognosis is so bad that a group of twenty-one young people in the United States is suing the federal government for the failure of its climate and energy policies to protect their habitat: Earth.

It is easy to find sport and intrigue in assessing Tesla's impact purely according to business metrics. Movements in the company's stock price, or mythic Musk's day-to-day pronouncements, are media manna, and there are always click-worthy headlines to be crafted based on proclamations that Tesla is on the precipice of catastrophic failure or the verge of infinite global dominance. And there is little doubt that, no matter what happens to its balance sheet from here, Tesla will continue to provoke extremes of pique and praise. But there can also be no doubt that in the stories told by my son's generation about humanity's success or failure to shift to sustainable energy, Elon Musk and Tesla will be leading characters.

Even if it died tomorrow, Tesla has already achieved what it set out to do: accelerate the world's transition to sustainable transport. It has convinced the world that electric cars can be great. It has overcome long-standing objections relating to range anxiety, insufficient infrastructure, and cost concern. It has shown traditional automakers that they must move more aggressively into electric propulsion, and it has inspired a new generation of entrepreneurs who see opportunity in building on the work Musk's company has done.

And if Tesla doesn't die soon? Well, it has a shot at becoming that trillion-dollar company. In its second-quarter earnings report of 2017, it revealed that it had installed its first solar roofs, made up of shingles that contain embedded solar cells. It continues to build industrial-scale energy storage systems to reduce the need for fossil-fuel-based storage facilities and peaking power plants, which typically sit idle but are used at times of high electricity demand. And the Model 3, the most important vehicle in the history of electric cars, garnered close to half a million preorders before the first one had even been delivered. That number puts it in the same bracket as BMW's comparably priced 3 Series, the German automaker's bestselling car for decades.

The Model 3 is to the 3 Series what the iPhone was to the BlackBerry. There's no key for the Model 3, just a Bluetooth connection through a smartphone that unlocks the door when you approach. There are no knobs or buttons on the dashboard. All of the car's controls reside in a fifteen-inch touch screen mounted horizontally above the center console. The air-conditioning emanates from a single continuous vent that runs the length of the dashboard and can be diverted to various parts of the cabin by dragging a finger around the screen. A full glass roof offers infinite views of the sky. Software can do most of the driving.

As 2017 drew to a close, I had a chance to drive the car, not long after I had test-driven the Chevy Bolt. There was no comparison. I had spent plenty of time in a Model S and a Model X, but sitting behind the wheel of the Model 3 immediately gave me the sense that I was in a next-level family car. It felt like a smaller version of the Model S, which is a good thing because the larger sedan verges on too wide and muscular for my liking. But it still had plenty of room. At one point, we packed in four adults and a baby in a child seat, with no complaints from any of the occupants. The power underfoot still felt immense and instantaneous, even though its zero-to-sixty time was closer to five seconds than the Model S's three. And the cabin, free from the clutter characteristic of normal cars, felt like something designed for today's consumer. It was the closest thing to an Apple product that exists on four wheels.

Like the iPhone, the Model 3 is more expensive than what many of its owners would have paid for a product in its class before it came along. But, also like the iPhone, it is so different from what has come before that people get the sense that they're buying more than just a product—they're buying magic. If Tesla can make the Model 3 in anything like the numbers it has promised, the car stands to bring the company what the iPhone brought Apple: an explosion of sales, and a product that reshapes an industry.

One night in the summer of 2017, my wife and I were driving our 2001 Honda when we were sideswiped by a Volkswagen that had butted into our lane. The damage was slight, but it was expensive enough that the insurance company decided that the old vehicle was a write-off. We considered replacing it with a Honda Fit or Toyota Prius but ultimately decided to go carless for at least a few months. We had placed a reservation for a Model 3 and hoped that our financial situation would improve to the point where we could pay for the car when it became available—perhaps in early 2018.

It's my hope that the Model 3 will be the family car that our son will grow up with. We have no intention of ever again buying a gasoline vehicle, which means he could experience transport powered by the internal combustion engine only in rare circumstances. In each case, it will seem to him like stepping back into another era, like kids today encountering rotary-dial telephones. By the time he is an adult, I hope, he will look back on this time and wonder why anyone got so worked up about oil, and how anyone doubted its period of dominance could ever end.

14

CATCHING A RIDE
TO THE RENAISSANCE

"The trends here are irreversible."

I sat in the back of a white Volkswagen Jetta beside Gansha Wu as an Uber driver took us from a hotel near the Beijing international airport to the tech business district Zhongguancun during morning rush hour. Cars sidled up beside us and then shunted their noses in front. They came streaming around corners without slowing. Sudden evasive action was just another defensive driving maneuver on these overburdened streets. We were just a bug in a swarm of metal and fumes.

The average driving speed in Beijing is 7.5 miles an hour.

As we stopped at a traffic light, Wu started talking about economic revolutions. He had read *The Major Economic Cycles*, a 1925 book by the Soviet economist Nikolai Kondratiev, which controversially argued that technological revolutions coincided with economic cycles, each of greater import than the last. Stalin had Kondratiev shot for suggesting that anything other than government had control of the economy.

The light changed and the cars around us moved slowly forward. An impertinent honk came from somewhere behind us. Wu had an epiphany while reading Kondratiev's book. The world had already gone through mechanical, electrical, and high-tech revolutions, he observed, and the last wave ended with the economic crisis of 2008. Now, he believed, it was entering an artificial intelligence revolution that could have a profound and lasting effect on humanity. He saw an opportunity. "We are at the beginning of a new cycle."

Wu owned a Volvo V60, but he couldn't drive it that day, a Thursday in May 2016, because its license plate ended in the number 5. The Beijing government restricted personal vehicle use so that, one day a week, each car took an enforced break. We sat in the back of the Jetta. Wu was in a shirt and jeans and had his backpack on his lap. I had my backpack on the floor between my legs. He wore dark sneakers, had tidy salt-and-pepper hair, and in his right hand he clutched a large Samsung smartphone. He was, at forty years old, one of a new generation of Chinese entrepreneurs and had chosen to enter a difficult industry: automotive.

Until 2014, Wu had spent his entire career at Intel Labs' Beijing office, where, after fourteen years, he became managing director. He grew up in the city of Haimen, home to a million people across the Yangtze River from Shanghai, and became interested in math at a young age, taking part in regional math competitions. At seventeen, he started studying computer science at Shanghai's Fudan University. His mother was a math teacher and his father, who had tried to start his own business on multiple occasions, had ended up in education, too. His last job was teaching aquaculture.

While at Intel, Wu was asked to write the introduction for the Chinese-language version of Michael Malone's book *The Intel Trinity*, a history of the company as told through the lives of its three most prominent leaders: Robert Noyce, Gordon Moore, and Andy Grove. During his research, Wu watched a webcast of Malone delivering a speech to an auditorium filled with hundreds of employees. Wu was struck by the writer's message to the crowd. "Intel is a company of destiny," Wu recalled Malone saying. "But that destiny can only be yours if you continue to take risks." The future, Malone told the crowd, was theirs for the taking. "If you are too careful, you will fail."

Reflecting on his own life, Wu concluded that he had become too conservative. Why had he stayed at one company for so long? His father, who was never successful in his multiple attempts to start his own companies, had taught him to be unafraid of failure. This was an unusual message for a Chinese parent to deliver in a society that tends to prize conventional career paths and job security. Wu saw that his father, whom he much admired, had been happy despite the disappointments in business. From this, he took strength. "Failure is nothing," Wu said he learned from his dad. "If you have your dream, if you have your faith, failure cannot beat you."

Wu spoke carefully, always supporting his observations with numbers. It was he who told me the average driving speed in Beijing. He also listed a series of other facts. There were about six million cars in Beijing, and more than two million people waiting to have their names drawn in the license plate lottery. (Only 150,000 new cars were licensed in Beijing that year.) Wu had taught himself English during his time at Intel, where he frequently interacted with American colleagues and made business trips to the United States. He was growing increasingly interested in artificial intelligence at the time he saw Malone's speech, in part because of a friend named Yong Zhao. Yong had been a founding member of the team that worked on Google Glass, the augmented-reality headset that overlaid a digital interface onto the real world (you may recall that the device looked like a pair of lensless sunglasses from the 2052 Olympics). After leaving Google in 2013, Yong started an automobile-vision research company but later decided he wanted to pursue other interests. He was looking for someone to start a company that would put his research to use.

As we crept past a tree-lined park, Wu said the auto industry was ripe for reformatting. "In the last century, the automobile industry actually hasn't changed a lot," he said. Two people on a moped, no helmets, swept by. He had read that there were more than 250 auto companies worldwide in the early 1900s but that there were now only about fourteen (depending on how you count the consolidated entities). Electric vehicles, however, had "changed the whole situation." The number of components in an automobile could be drastically reduced, and the value chain was changing significantly. New competitors had a better chance than ever.

For the last century, Wu said, the auto industry had been like boxing, dominated by big guys who followed well-defined rules. But now, it was like mixed martial arts, which favors scrappy, nimble fighters, even if they're smaller. "The rules become very complicated—or people don't follow rules." Meanwhile, ride-sharing had become popular, led in China by Didi Chuxing. Connectivity provided an opportunity to make cars more like mobile phones. And the idea of autonomous driving was gaining momentum, with Baidu, China's largest online search company, working on its own vehicle since 2013, following Google's lead.

I started to get a little nauseous. It wasn't only because I was focusing my attention on Wu instead of the road but also because the driver, like many in China, was unkind to the brakes. It was 77 degrees Fahrenheit outside, and we had the front passenger window open. My breakfast was a cocktail of car fumes and industrial smog. As we advanced farther into the city, the buildings around us rose taller, traffic slowed, and the smog thickened.

Wu had a vision for the future that would clear all this up—the smog, the crowded roads, the rough driving. Five years from now, he said, there would be two types of cars in China. One would be a high-speed passenger car that emphasized safety. The other would be a low-speed urban commuter, not sold to consumers but available on demand via a smartphone app. These "robotaxis" would have a top speed of forty miles an hour and would be able to communicate with each other to drastically improve traffic flow. In ten years, we would see millions of such robotaxis on the road, Wu predicted. A Massachusetts Institute of Technology study had shown that ride-sharing and car-sharing could reduce the number of cars on the road by 80 percent. He believed that there would be three million cars in Beijing by 2026—half as many as there were in 2016—of which two million would be robotaxis. As I talked to Wu, there were only about seventy thousand taxis in the city for a population of twenty million people. It was hard to get a ride.

Wu pulled a fifteen-inch PowerBook from his bag. With the computer perched on his lap, he brought up a video animation that showed autonomous cars—represented as little white rectangles against a black backdrop—flowing steadily through an intersection without the need for traffic lights. The hypothetical cars were communicating with each other so they knew when to slow down, when to yield for others, and when to speed up again. In another video, Wu showed how sudden braking causes traffic jams. When a white rectangle on the screen suddenly halted its forward motion, it sent a shock wave out its backside, affecting dozens of vehicles behind it, each having to brake in turn. With computers as drivers, autonomous cars would be better at monitoring optimum speeds and adjusting smoothly to the conditions without having to use the brakes as often.

On an ordinary workday, Wu leaves his apartment on the outskirts of Beijing at 7:40 A.M., walks fifteen minutes to a subway station, and takes a train to Zhongguancun, where his autonomous driving start-up, Uisee, is based. It's a forty-five-minute ride, during which he reads business and technology books. (He had been reading the *New York Times* reporter John Markoff's history of AI, *Machines of Loving Grace*.) After getting off the train, Wu takes his bike out of a locker and rides the remaining twenty minutes to Uisee's office, which, when

I visited, was in a shared working space on the fifteenth floor of a building that once housed stores selling cheap electronics. The incubator, called My Dream Plus, looked like it could be in Silicon Valley, with colorful bucket chairs and brightly upholstered ottomans interspersed with potted plants and interior walls adorned with ivy. Companies were segmented into rooms with glass doors, each large enough to hold about thirty people. Uisee had twenty-two employees in this office and a team of automotive engineers in Shanghai. It was working with a small independent manufacturer to test its software in real vehicles. A year later, it had 150 employees at three offices in Beijing and Shanghai, with plans for a fourth in Shenzhen in 2018.

Uisee is an acronym that stands for Utilization, Indiscriminate, Safety, Efficiency, and Environment. The company is working on autonomous driving systems based on a supercomputer that communicates with a combination of cameras, millimeter-wave radars, ultrasonic radars, GPS, and inertia units that can keep track of the car when the GPS doesn't work—for instance, when signals from satellites are blocked by tall buildings. Wu and his cohorts hadn't yet come up with a crisp mission statement for the company, but they knew they wanted to make transportation enjoyable and safe. "Everyone deserves a personal driver backed by AI," said Wu. For the first applications of its technology, Uisee trialed self-driving vehicles for cargo and passengers at the international airports in Singapore and Guangzhou, in southern China.

China could certainly use self-driving cars. More than seven hundred people are killed on its roads a day, according to the World Health Organization. In a phone interview two days before I met Wu, Wang Jing, the head of Baidu's autonomous driving unit, said Baidu believed self-driving cars could reduce the death toll on China's roads by 90 percent—since that's the proportion of accidents caused by human error. Self-driving cars would also save time, Wang said. A commute in a megacity like Beijing or Shanghai commonly takes an hour or two, but most cars on the road have only one occupant. A more efficient distribution of bodies per vehicle through ride-sharing would improve traffic and speed up commutes.

In China, it will be easier than in the United States for people to make the transition to sharing self-driving cars, Wang reasoned. The population density of the big cities makes it difficult for people to have access to their own parking spaces, and hundreds of millions of Chinese still don't own cars. "For Chinese people, they don't have this concept; they haven't completely embraced the private car yet," Wang said. "So it's easier for them to embrace the autonomous-driving car with the car-sharing." He also argued that China provides the ideal testing conditions for autonomous vehicles, because the roads are so crowded and driving behaviors are so erratic, presenting a panoply of challenges for the cars' supercomputers to contend with—and ultimately overcome. "Technically, if you can make it work in China, you can make it work anywhere."

A day before my call with Wang, Baidu had announced a partnership with Wuhu, a city of nearly four million people in Anhui province, to start a self-driving-car pilot program in the central city. For the first three years, self-driving cars, vans, and buses would be introduced to downtown areas purely for testing. Beyond three years, the plan is to start commercializing the program and allowing passengers in the vehicles. Ultimately, the program would spread across the whole city. Wang cited Wuhu as an example of how regulation in China might actually work in favor of autonomous vehicles. The city government was "very enthusiastic" about the project, he said. And besides, even if governments were slow to embrace the technology, the advent of self-driving cars was inevitable. "That's the mega trend. It's definitely coming."

In Wang's opinion, software will be the most important car spec of the future. Today, we obsess over a car's engine size, acceleration time, or fuel economy—but that's going to change. In the future, we'll want to know whether our car can communicate with the traffic system, other cars, or even the road itself. We'll want to know how far ahead our car can "see," so we can assess how safe it is. We'll want to know what its driving patterns are like, how smooth a ride it can offer. These concepts, Wang reasoned, are similar to how we assess smartphones today. The software that the phone is running—iOS or Android—matters more to many consumers than the phone's hardware or central processing unit.

In the United States, Tesla's JB Straubel shares Wang's view. "Autonomous vehicles are mostly defined by the software that operates them," Straubel said at a conference in 2016. The improvements happening in the field were phenomenal. "We can literally measure them in months instead of years." Tesla was paying special attention to image recognition, Straubel said. "The trends here are irreversible. We're not going to see them slow down or stop."

In the summer of 2016, a friend picked me up for a ride in a Tesla Model X, the first time I had been in the car. On the highway just south of San Francisco, he put it into Autopilot mode and took his hands off the steering wheel. Even though I had read all about the technology and had seen it on video, I still instinctively flinched. As soon as he did it, I wanted to reach over and guide the car myself. I was perturbed that, even as cars sped along beside us, the Model X did not slow down. If anything, it felt like it sped up.

The car drove itself at the speed limit, sixty-five miles an hour, and stayed glued to its lane with equidistant clearance on either side. It took about twenty seconds for my breathing to return to normal. Still, I kept one eye on the road and one on the wheel. Then, just as I was calming down, my friend flipped on the turn signal and the car quickly, but smoothly, guided itself into the lane to the right, maintaining a safe distance from other cars.

"Do you trust it?" I asked my friend.

"Yeah, I do," he said, with little hesitation.

He'd driven with Autopilot many times by then. It didn't seem like a big deal to him anymore. "It's especially useful in stop-and-go traffic when I'm commuting. It's such a big help in reducing the tedium, and not having to think about what my hands and feet are doing every second."

Our AI-guided drive quickly got boring. We were soon talking to each other about other things, and the car's activity became background. I didn't feel at risk, and I didn't feel safer when, after a couple of minutes, my friend retook the wheel. It was not difficult to imagine that the concept of a car that drives itself will one day seem unremarkable.

At the time, Tesla's Autopilot system was classified as Level 2 autonomy, meaning the car could keep itself in a lane, change lanes, and modulate speed according to traffic conditions, but that a human driver still had to be present, attentive, and ready to take over at any moment. The National Highway Traffic Safety Administration had established a framework for categorization that was based on six levels. Level 0 was the lowest, meaning that a human had complete control with no computer assistance, while for Level 1 the car had some advanced driver-assistance technology, such as automatic emergency braking, but the driver still controlled the vehicle at all times. Level 5 was the highest, at which a car would have no controls for human drivers whatsoever. At that point, you could read a book, take a nap, or watch a movie while the car drove itself. Google has tested fully autonomous vehicles to a Level 5 designation, meaning the cars could perform all "safety-critical driving functions and monitor roadway conditions for an entire trip," but they haven't yet left the test circuit.

The development of autonomous vehicles goes hand in hand with the development of electric vehicles, because self-driving cars are best controlled by drive-by-wire systems, in which electrical signals and digital controls, rather than mechanical functions, operate a car's core systems, such as steering, acceleration, and braking. The absence of a large engine block, too, opens up more design possibilities for electric vehicles, so autonomous cars could come in more varied shapes and sizes, as small as a covered Segway or as large as a double-decker bus.

But to the extent that the spread of autonomous vehicles depends on electric vehicles, so, too, must they depend on the expansion of electric vehicle infrastructure, especially the proliferation of charging stations. And who wouldn't want a car that could refuel itself? Electric cars make this relatively straightforward, either through self-charging mechanisms—Tesla has built a prototypical metallic "snake" that finds its own way to a charge port—or wireless charging embedded in roads or pads.

Ford has created a subsidiary called Ford Smart Mobility that is focusing directly on this challenge. "The point we've been making is that it's not moving from an old business to a new business," Ford's then CEO Mark Fields said in an April 2016 interview with *The Verge*. "It's just moving to a bigger business as we expand the business model from number of units sold to number of units sold plus vehicle miles traveled." In August 2016, Ford announced plans to bring a Level 4 self-driving car—without pedals or a steering wheel—to market by 2021.

Other automakers have been working on similarly aggressive plans. Fiat Chrysler has partnered with Google's Waymo to develop a fleet of self-driving hybrid minivans. GM, through its partnership with Lyft, has plans to bring Chevy Bolt robotaxis to the road as quickly as possible. Mercedes has been so eager to tout its semiautonomous features that it got into trouble for promoting its E-Class model as "self-driving" (in fact, the car had only Level 2 autonomy and was compared unfavorably to Tesla by the auto journalist Alex Roy). It was forced to withdraw the ads. Volkswagen has made self-driving cars an important part of its ten-year plan in the wake of the diesel emissions scandal that caused it to rethink its entire strategy. Many major auto-makers have established research centers in Silicon Valley to work on autonomy, including Nissan, Toyota, Mercedes, Ford, and GM. The newcomers—Apple, Lucid Motors, Faraday Future, Byton, and Nio—have made autonomy central to their business models and es-tablished software development teams in California. Che He Jia and Singulato Motors are working on the technology in Beijing and Shanghai. In the meantime, other tech companies and start-ups, such as Uber, Lyft, Comma.ai, Nauto, Luminar, Aurora, Caracal, Starsky Robotics, and Zoox, are all chasing variations of the self-driving prize, be it for cars, buses, or trucks.

It is probably safe to say, as Gansha Wu did, that cars that can drive themselves at low speeds in controlled environments, such as inner cities, are likely to be available within ten years, and perhaps even as soon as 2018. These cars will start off on campuses, at airports, in theme parks, and in special urban test zones, such as the one in Wuhu, China. In July 2016, a self-driving bus made by Mercedes navigated a 12.5-mile journey through tunnels, bends, and traffic lights from Amsterdam's Schiphol Airport to the town of Haarlem. Self-driving golf carts have already started roaming university cam-puses in California, and in August 2016 a company called nuTonomy started testing a free self-driving taxi service in a small business dis-trict in Singapore.

But full autonomy for high-speed cars could take much longer. At the South by Southwest tech conference in March 2016, Chris Urm-son, who was then the head of Google's self-driving car program, said that in some places autonomous vehicles won't be on the roads for as many as thirty years. At the same time, Elon Musk, ever the optimist, has said that he thinks Tesla's cars will be ready for "complete auton-omy" by 2018, but that the regulatory process will add another year to the rollout. In October 2016, Tesla said that all its new cars would be equipped with hardware that would allow full self-driving at some point in the future. The company said it would "further calibrate the system using millions of miles of real-world driving." A Morgan Stanley analyst has predicted that complete autonomy will be ready by 2022, with massive market penetration coming by 2026.

In bad weather, it's harder for an autonomous car's cameras to make out markings on the road and see other vehicles in front of them. Snow, ice, or mud can cover the cameras, blocking their views. Many autonomous-driving systems depend, to a large extent, on high-precision 3-D maps, which can be accurate to within inches and help cars anticipate corners, hills, and exit ramps. However, such maps will take time to build out and they age quickly as road conditions change and infrastructure develops. Experts have suggested that universal and standardized mapping data that can be shared across vehicle fleets and updated in real time will be necessary for the self-driving era, but you can bet that it will take some time for the various stakeholders to agree on a coherent plan for such cooperation. Cars will also need steady, reliable, and fast wireless connections so they can communicate with cloud servers, other vehicles, and infrastructure such as traffic systems and parking structures, even in remote areas. That effort will be helped by the advent of the 5G wireless standard, but that likely won't be commercially available until well after 2018, which is when Intel planned to start some 5G trials.

Regulation will also likely slow down the technology's proliferation. As the crash that killed Joshua Brown, the first Autopilot fatality, showed, many people are still unsure about the safety of autonomous vehicles—and subsequent deaths related to self-driving cars have only increased the concern. *Consumer Reports* wrote that Tesla had pushed autonomy too far, too soon, and called on it to disable automatic steering and require drivers to keep their hands on the wheel.

A key regulatory issue will center on the "handoff problem." Evidence suggests that Level 2 or 3 systems like Autopilot may engender a false sense of security among human drivers, who are prone to stop paying attention for periods during which the cars steer themselves. A small study by Virginia Tech tested the fortitude of twelve drivers who were sent on a three-hour semiautonomous drive around a test track and tempted with videos, magazines, books, and food. Three of the drivers used their hands-free driving time to read, and seven took the opportunity to watch a DVD.

Another important debate will center on the ethical decisions that the cars' computers must make in emergency situations. Should a car choose to hit a pedestrian in order to save the life of its occupant? Or, faced with a choice between crashing into two elderly citizens on the left side of the road or one infant on the right, what should an autonomous car do? What if, in a case where the system determines that fatalities are certain to occur, a car's computer knows it can ensure a victim has a more humane death if it accelerates toward him? Answering such questions could keep regulatory debates running in circles for years.

Lobby groups will keep the regulators busy. In April 2016, as NHTSA held public hearings about self-driving cars, a group that included Ford, Google, Uber, Lyft, and Volvo announced the formation of the Self-Driving Coalition for Safer Streets, led by David Strickland, a former NHTSA administrator. The group has been advocating for a clear set of federal standards for autonomous vehicles in the United States. In June 2016, the National Association of City Transportation Officials, a coalition of officials from dozens of large North American cities, published a policy statement that included a series of safety- and civic-minded recommendations, such as capping inner-city speeds for autonomous vehicles at twenty-five miles an hour and offering federal and state incentives to cities that prioritize self-driving electric cars that can be shared. And that September, the Obama administration issued guidelines that covered how driverless cars should behave in the case of a system failure, how they should preserve digital security, how cars should communicate with passengers, and how occupants should be protected in crashes. While unveiling the guidelines, the head of the National Economic Council said that highly automated vehicles will "save time, money, and lives."

In the meantime, research continues apace. Automakers and suppliers are testing autonomous vehicles at the University of Michigan's "Mcity" facility, a $10 million mock urban and suburban environment in Ann Arbor. Honda has been testing self-driving vehicles at a former naval munitions site called GoMentum Station, which features twenty miles of roads, tunnels, and other infrastructure in Concord, California, about thirty miles northeast of San Francisco. In Shanghai's Auto City, international and domestic automakers are putting self-driving cars through their paces in a test zone that will expand to thirty-nine square miles, including highways, within five years.

In April 2016, a convoy of a dozen trucks more or less drove themselves across Europe. The demonstration, sponsored by the Dutch government, involved trucks made by Scania, Daimler, Volvo, MAN, Iveco, and DAF, which negotiated the highways through Sweden, Germany, and Belgium to the Netherlands without the drivers needing to touch the wheels or pedals. While not fully autonomous, the trucks were connected to each other wirelessly and used automatic cruise control to "platoon" together, meaning each truck could trail less than a second behind another—a smaller gap than would be advisable if humans were in charge. Through platooning, trucks can benefit from drafting, meaning they are protected from the resistant effects of the wind by the vehicles in front of them. Platooning results in an average fuel savings of 10 percent per truck, according to a study by the Dutch research group TNO. The group calculated that two trucks platooning for a hundred thousand kilometers (about sixty-two thousand miles) would save €6,000 (about US$6,500) a year on fuel compared to driving in normal cruise control.

The European convoy is a taste of what's to come. A start-up called Convoy—backed by Amazon founder Jeff Bezos and other tech luminaries—is attempting to apply Uber's logistics-on-demand model to trucking, with a focus on short-haul trips. Shipping software start-up Flexport, backed by Google Ventures, wants to be the "Uber of the oceans." There's Tesla, of course, with its autonomous electric semi-truck, and other electric truck start-ups Nikola, Thor, and Starsky Robotics. A crack team of engineers from Google's self-driving car team left the company to establish the San Francisco–based Otto, which said in August 2016 that it was moving "with urgency" to get commercially ready autonomous trucks on the road within two years. Two days later, Uber announced that it had acquired Otto. The company's cofounder Anthony Levandowski—a pioneering engineer on Google's self-driving car team—would head up the ride-sharing company's autonomous vehicle efforts, and Otto would also lead Uber's efforts in trucking. That October, a self-driving truck controlled by Otto's technology and sponsored by Budweiser delivered a load of beer from Fort Collins, Colorado, to Colorado Springs—a 120-mile trip on Interstate 25. In February 2017, Google's Waymo filed a lawsuit against Uber, claiming patent infringement and alleging that Levandowski stole trade secrets, a dispute that led to one of the highest-profile court battles in tech history. The two companies ultimately reached a settlement agreement, with Waymo receiving 34 percent of Uber's stock.

Uber also revealed that it had partnered with Volvo for a self-driving car program based in Pittsburgh, Pennsylvania, starting with a trial run that would offer free rides in autonomous Ubers in the city (although each car would have a human in the driver's seat, just in case). Uber said it would have fully autonomous vehicles on the road by 2021. This target was thrown into doubt when, in 2018, one of its self-driving vehicles killed a pedestrian while on a test drive in Phoenix, Arizona.

Jobs will be lost. There are 3.5 million professional truck drivers in the United States. As soon as trucking companies can increase their bottom lines by cutting their labor costs, those drivers will find themselves no longer needed. But that won't be all. There are 5.2 million people in the trucking industry who don't drive trucks, as well as the millions who provide food, gasoline, accommodation, and other services for truckers. If truck drivers are no longer on the roads, all those people will feel the pain, too. Then you can look at the people who drive taxis, Ubers, and Lyfts. Many taxi drivers have already switched to driving for the ride-sharing companies, but when robotaxis and self-driving Ubers are widespread, many of those jobs will be at risk.

Some observers believe that the advent of the autonomous era could have a measurable impact on capitalism as we know it. Revenue from fuel taxes will go down, presumably to be replaced by other sources of income. Parking revenue—including fines—may well all but disappear. Speeding tickets and driver registrations will be greatly reduced. These developments are going to affect how governments make money and citizens spend it. Robin Chase, the former CEO of car-sharing company Zipcar, and now the executive chairman of vehicle-communications company Veniam, has called for a universal basic income to offset the losses that will be brought on by an era of automation. Such guaranteed income would allow "more people the opportunity to focus on purposeful, passion-driven work," she wrote in 2016. Instead of taxing labor, Chase argued, it would make more sense to tax the technical platforms that generate the profits and "the wealth of the small number of talented and lucky people who founded and financed these new jobless wonders."

Scientists at the University of California, Berkeley, have estimated that electric robotaxis could reduce greenhouse gas emissions by 90 percent compared to personal gasoline cars. A 2015 study by scholars at the University of Texas found that one self-driving car could replace nine regular vehicles. Cars that don't have to "cruise" for parking spots also release fewer emissions. In one study from 2007, an urban-planning professor at the University of California, Los Angeles, found that cars looking for parking in one Los Angeles business area generated 730 tons of carbon dioxide a year.

Columbia University's Earth Institute has found that shared autonomous cars would cost about fifteen cents per mile to operate, compared to sixty cents a mile for personal gasoline cars. Savings would come from improved driving efficiency, reduced wear and tear, and cheaper fuel (electricity). If you take the human driver out of the equation for Uber and Lyft, the cost of getting from A to B on four wheels would also be steeply reduced. We could sell our cars then, and never again have to worry about taking them in for inspection, changing the tires, or being exploited by unscrupulous mechanics. (Unfortunately, of course, people in those job sectors would see their incomes suffer.)

If we don't have to concentrate on the road during our commutes, we'll have more time for work and leisure. Maybe we'll watch more TV, read more books, do more crossword puzzles, and finally finish knitting that scarf. (More likely, though, we'll just reply to more work e-mails.) Meanwhile, because the computer will be better at driving than we are, we'll get into fewer accidents, and see fewer, too. Maybe our lives will be among the many thousands spared annually by having computers instead of humans control motor vehicles—an algorithmic miracle for which we'll never know to be thankful. Except for those among us who are especially twisted, we'll enjoy never again having to parallel park, or having to find parking at all, because our cars, or the ones we've paid to ride in, will drive themselves to parking spots or simply move on to pick up someone else. We'll be able to convert our garages into bedrooms or home offices, or places to store disused KitchenAid mixers, bread makers, and Abdominizers. We'll pay lower auto insurance premiums.

The autonomous age also promises a do-over for urban planning, and it might not be all to the good. People may forsake public transport and instead pour onto the roads, clogging them up even worse than today. Or, knowing that it's cheaper for their cars to simply drive themselves around the city than to pay for parking, car owners could let their empty vehicles roam while not in use, creating a nightmarish scenario of what Robin Chase calls "zombie cars" crowding the streets uselessly. Perhaps the suburbs will become so attractive as commutes become easier that instead of resource-efficient, high-density cities, we'll have endless urban sprawl. But it doesn't have to be that way.

Almost everything I know about the Renaissance, the period in European history from the fourteenth to the seventeenth century, comes from an eighteen-minute YouTube video produced by the author and philosophy guru Alain de Botton's School of Life. De Botton dedicates a few minutes of the video to educating viewers about the Renaissance leaders' zeal for building beautiful cities. You can count on one hand the number of cities built since the 1600s that can rival the elegance of cities that sprung up on the Italian Peninsula during the three-hundred-odd years of the Renaissance, de Botton says in the video. Sure, he concedes, the old urban planners didn't have to worry about cars or zoning laws, but they had a mission and were extremely direct and didactic in carrying it out. "City fathers across the Italian Peninsula had fallen in love with a remarkable new idea: that their cities should be the focus of an unparalleled attention to beauty," he says. "It's slightly embarrassing to contrast these efforts with our own."

De Botton argues that successful city planning is never an accident. The Renaissance produced great cities because its leaders believed that people are to a large extent shaped by the buildings around them. I am reminded of this comment when I walk the streets of San Francisco, a boomtown that ranks tenth in the world for number of billionaires, and must carefully avoid schmears of human excrement on the sidewalk. "Making sure that the public realm conveys dignity and calm is more than a luxury," de Botton says. "It can help to ensure the sanity, vigor, and happiness of a whole population."

The Renaissance leaders embraced the idea that the public sphere should be beautiful, refined, and appealing so that a society's richer citizens would never be tempted to withdraw into their private estates, closed off from the world around them. All citizens could then be "uplifted by a pleasing vision of communal life."

It was 1908 when Henry Ford unveiled the first Model T, a product that would reorient the infrastructure of civilization, and around which civilization would reorient itself. Just over a century later, Elon Musk unveiled the Model S at a time when civilization is more than ready for a cultural rebirth—one that could be catalyzed by something as innocuous as a beautiful car that drives itself. Autonomy, after all, is a term not limited to the automatic control of a motor vehicle. Its meaning also speaks of self-determination. It is through the power of this autonomy that we can turn a revolution into a renaissance.

A NOTE ON SOURCES

This book relies on a combination of my own reporting and that done by others. For all Tesla-related content, I have relied only on publicly available sources, including news stories, magazine profiles, blog posts, videos, documentaries, court documents, and company filings. For the sections about Tesla's and Elon Musk's histories, I am indebted especially to Ashlee Vance's biography, *Elon Musk: Tesla, SpaceX, and the Quest for a Fantastic Future*, Drake Baer's reporting for *Business Insider*, and Max Chafkin's 2007 profile of Musk for *Inc.* Chris Paine's documentaries, *Who Killed the Electric Car?* and *Revenge of the Electric Car*, were also valuable sources of material. For any non-Tesla content, I have relied chiefly on my own interviews and research, as should be obvious in the text.

I have compiled an extensive list of sources for the facts documented in this book and published them on my website, hamishmckenzie .com. I encourage anyone interested in the veracity of my reporting to check every detail as listed there.

I relied on my memory to reconstruct conversation for some instances of dialogue—in all cases, minor—but I transcribed the vast majority of dialogue from recordings of my own interviews or from videos available online. On a few occasions, I have reconstructed scenes from video, photos, and media reports.

I also want to take a moment to pay tribute to the many underpaid and underappreciated reporters whose work has provided the bedrock for understanding many of the subjects and personalities covered here. My thanks in particular go to the reporters at *The New York Times*, *Bloomberg*, *Business Insider*, *Businessweek*, *The Wall Street Journal*, Reuters, *Fortune*, *Inc.*, *Wired*, *The Guardian*, the *Los Angeles Times*, the *Huffington Post*, *Handelsblatt*, *South China Morning Post*, and *Caixin*, without whom a work of this nature wouldn't be possible.

ACKNOWLEDGMENTS

Insane Mode was germinated in August 2013, when Julian Loose, then head of nonfiction at Faber & Faber, e-mailed to ask if I'd be interested in writing a book about Elon Musk. Julian has since provided much-needed emotional and professional support as the project that he set in motion sprinted, then stumbled, and finally ambled to its conclusion. I'm not sure how I can adequately thank him for his help, but I hope these lines go some way to doing so.

My agent, Jim Levine, was an early believer who had more confidence in my abilities than I possessed myself. I have been lucky to have him on my side and to benefit from the wisdom of his experience. I also extend thanks to Jim's colleague Beth Fisher, who attracted the interest of my international publishers, and to everyone else at the Levine, Greenberg, Rostan Literary Agency for their help and patience.

Getting to work with my editor, Stephen Morrow, a fellow small-town antipodean, was also an immense stroke of luck. Stephen seemed to get me on every level and made incisive suggestions that dramatically improved the book while letting me hold on to the voice and values that I care about most in my writing. He also supported me through some unexpected challenges. I am immensely grateful.

Thanks, too, to Angus Cargill, my editor at Faber & Faber, for his advice and guidance.

Thanks to Scott Gattey for his legal advice; to Victor Menotti for his research support; and to Hans Tung for connecting me to entrepreneurs in China.

Martin Leach, the cofounder of NextEV and forty-year veteran of the auto industry, was generous with his time and forthcoming in our interviews. I was saddened to hear of his passing in November 2016. I hope I have represented him well in these pages.

I owe a lot to Shane Snow, a great guy and multitalented writer-entrepreneur, who opened doors to which I could otherwise never have hoped to get close. Thanks to friends Kelly Pendergrast, Chris House, Patrick Crewdson, Murdoch Stephens, Christopher Mims, Ashlee Vance, Cullen Thomas, Liz Jarvis-Shean, Alexis Georgeson, and Christina Ra—and others too numerous to name—who have all provided immense moral support. I am also grateful to my former colleagues and current friends at Tesla and Kik.

I owe Ted Livingston, Kik's CEO, deep gratitude. I am a beneficiary of Ted's relentlessly optimistic quest for the win-win. He's a tremendous guy and a role model. So is Chris Best, Ted's former cofounder, and now my cofounder at Substack, our subscription publishing start-up.

To my brother, Andrew, and his wife, Miriama, I say a big fat *kia ora* and thank you for your encouragement and sympathy when I needed it. Your kids are cool, too. (Hi, Tommy, Rachel, and Esther!) Also, *kia ora* to my other brother, David, who's not around anymore but who taught me so much while he was.

To my mum, Louise, thank you for your love and the guts to keep going when things got tough. And to my dad, Richard, thanks for dutifully reading every version of every chapter, and your sheer enthusiasm for this project.

To my son, James, who arrived on July 5, 2017—you may never know how much you helped me get through some hard times. Thank you.

Finally, to my wife, Steph Wang. Thank you for your unwavering support and help through every minute of this project. This is just the latest of the many journeys we have traveled together, and a precursor of many more to come. I hope the next one isn't quite so bumpy. I love you.

INDEX